the *Joy* of TINY HOUSE LIVING

the Joy of TINY HOUSE LIVING

CHRIS SCHAPDICK

Everything YOU NEED TO KNOW BEFORE TAKING THE PLUNGE

CREATIVE
HOMEOWNER®

CRE**A**TIVE
HOMEOWNER®

The Joy of Tiny House Living
Vice President-Content: Christopher Reggio
Editor: Colleen Dorsey
Copy Editor: Laura Taylor
Designer: David Fisk
Indexer: Jay Kreider

ISBN 978-1-58011-834-7

Library of Congress Cataloging-in-Publication Data

Names: Schapdick, Chris, author.
Title: The Joy of Tiny House Living / Chris Schapdick.
Description: Mount Joy : Creative Homeowner, [2019] | Includes
   bibliographical references and index.
Identifiers: LCCN 2018049952 | ISBN 9781580118347 (pbk. : alk. paper)
Subjects: LCSH: House construction--Amateurs' manuals. | Small
   houses--Maintenance and repair--Amateurs' manuals.
Classification: LCC TH4815 .S375 2019 | DDC 643/.2--dc23
LC record available at https://lccn.loc.gov/2018049952

We are always looking for talented authors. To submit an idea, please send a brief inquiry to acquisitions@foxchapelpublishing.com.

Printed in Singapore

Current Printing (last digit)
10 9 8 7 6 5 4 3 2 1

Creative Homeowner®, *www.creativehomeowner.com*, is an imprint of New Design Originals Corporation and distributed exclusively in North America by Fox Chapel Publishing Company, Inc., 800-457-9112, 903 Square Street, Mount Joy, PA 17552, and in the United Kingdom by Grantham Book Service, Trent Road, Grantham, Lincolnshire, NG31 7XQ.

Dedication

This book is dedicated to my parents and my daughter, Mia. My parents gave me the path and my daughter gave me the reason to pursue this fulfilling life of creating things with my hands. Thank you for all your support and love.

The man behind the man: my dad. Thanks for believing in me and for all your help.

All costs related to your tiny home project will vary by project, location, and many other variables; therefore, all costs given in this book should be understood to be estimates.

All interviews have been edited for length and clarity.

Is This Book for You?

You are tiny house curious. Going tiny appeals to something deep inside of you. Maybe you like to watch shows about tiny home living on TV. You want an escape from some of the monotony of daily life. You like the idea of reducing clutter and excess and are perhaps interested in minimalism.

If any of these statements are true about you, then tiny houses, structures, buses, or other small dwellings may well be the answer for you. This book describes the many aspects of going down the path toward your own tiny house. Whether you want to build it yourself or have someone build it for you, there are things you should know and decisions you will need to make. I'm here to arm you with the information you need to make those decisions.

This is not a step-by-step how-to book that teaches you to build a tiny house from scratch—though that will be my next book! Instead, when you are done reading this book, you'll be well equipped to make your own tiny house decisions. From design to construction to where you ultimately put your tiny home, I'll explore your options and help you find clarity. This book is laid out in a logical progression that your tiny house journey is likely to follow. I suggest that you read it through in its entirety before making big decisions so that nothing outlined later on in the book will trip you up.

I wholeheartedly encourage you to embrace smaller living. It's pretty groovy, and no, it's not just a fad.

Table of Contents

Foreword

DEREK "DEEK" DIEDRICKSEN

This was the first time we met in person. We needed to get a selfie.

Chris Schapdick is a pretty curious fellow. To be frank, he has just a little bit of that mad scientist look about him, a zany "Gene Wilder as Frankenstein" vibe. I intend that comment to be a compliment, though.

Chris has a gleam about him of an individual who is very bright and who has many onion-like layers. Chris, like myself, was probably a total nerd in high school, but it's those people that you want to align yourself with. The dorks of yesterday are the innovators, pioneers, and often the bosses (perhaps yours) of today. They are the ones who were never afraid to go against the grain, who have always done so, and who stick their necks out, take chances, and willingly sprint down the path less chosen. Heck, a lot of 'em, like Chris, grab proverbial machetes and blaze their very own paths. Quite honestly, you are cut from the same cloth if you've made the rather wise choice to pick up this tome.

Even adventurers need a bit of guidance sometimes, though, and that's what you'll find within the pages of this book. The sum of Chris's experience in the tiny house world, building and experimenting and learning, is collected here in easily digestible, thought-provoking, often entertaining and always useful chapters that cover everything from the philosophical reasons to go tiny to the pros and cons of different trailers and building materials…and everything in between.

When years back, Chris decided, "I'm going to start a business based on tiny houses and micro gypsy wagons, all from scratch!" it was not your normal nine-to-five by any means, and certainly a chancy, gutsy, move—yet Chris has done remarkably well, and there is no reason why you can't go down that path, too, by combining your own sense of adventure with the confidence that the facts and advice in this book will give you.

So read heartily and drink from this fountain of micro-architecture geekiness. Mr. S is your more-than-capable tour guide from here on out, and he'll be taking you through the journey of what tiny living is all about, why you might want to consider this path, and how to do it.

—Derek "Deek" Diedricksen

Author of *Microshelters* **and** *Micro Living*, **former host of HGTV's** *Tiny House Builders*, **and creator of YouTube channel relaxshacksDOTcom,** *www.youtube.com/user/relaxshacksDOTcom*

Gallery

This gallery of photos features real people and real tiny houses. You can read it word-for-word now, before you get into the meat of the book, or you can skip it and come back to it later, or you can skim it and flip back to it as you read about the ideas and concepts that are shown in the photos. No one's tiny house is perfect for another person, but by seeing what others have done with limited space, you'll be inspired to figure out what your tiny life will look like.

Courtesy of TinyHouseNC (www.tinyhousenc.com). Photo by Mandy Lea Photography.

Courtesy of TinyHouseNC (www.tinyhousenc.com). Photo by Mandy Lea Photography.

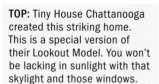

TOP: Tiny House Chattanooga created this striking home. This is a special version of their Lookout Model. You won't be lacking in sunlight with that skylight and those windows.

RIGHT: The bedroom area in the Lookout by Tiny House Chattanooga has its own dedicated skylight and breathing space on both sides of the bed—something to appreciate!

ABOVE: This is the Cardinal model by Free Spirit Tiny Homes. It's 20' (6m) long and offers 196 square feet (18 square meters) of living space.

RIGHT: An inside view of the Cardinal model by Free Spirit Tiny Homes reveals wooden walls that make it feel like a modern log cabin.

ABOVE: The inside of Artisan Josh's 12' (3.6m) tiny house. The dark wood and number of decorations add to the feeling of coziness here, and enough light gets in the windows to keep it from feeling like a cave. (www.artisanjosh.com)

TOP RIGHT: Bryan Booth from Harmony Tiny Homes stands in the Youngstown model, a 270 square foot (25 square meter) and 24' (7.3m) long tiny house. Bright colors enhance the open and spacious floor plan.

RIGHT: The Bluestem from Switchgrass Tiny Homes can fit quite a few people comfortably. Surfaces hide storage space, and utilities, like the mini split unit at top left, are out of the way when possible.

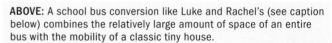

ABOVE: A school bus conversion like Luke and Rachel's (see caption below) combines the relatively large amount of space of an entire bus with the mobility of a classic tiny house.

LEFT: "Midwest Wanderers" Luke and Rachel (plus their two kids and dog) live in this amazing school bus conversion. Aside from living in a bus, Luke converts buses for others with his company, Skoolie. (www.skoolie.com)

BELOW: The exterior of the bus.

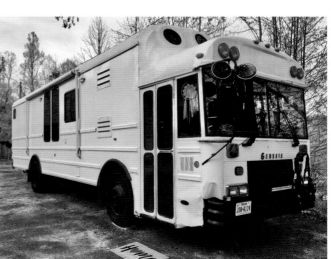

ABOVE: Alexis Stephens and Christian Parsons stand in front of their nomad home, where they've been living and traveling for three years and counting.

RIGHT: This is the view of the "greatroom" from the front door of the house. The uncluttered feel allows a small space to live up to its name.

Courtesy of Tiny House Expedition (www.tinyhouseexpedition.com).

Courtesy of Tiny House Expedition (www.tinyhouseexpedition.com).

TOP LEFT: Facing the front of the house, you can see a creative vertical shoe rack and a small extra bedroom area—no space is wasted!

TOP RIGHT: There is even space for proper utilities, like a mini fridge, stove, and oven, in this tiny home with a big heart.

LEFT: Thoughtful touches in a consistent color scheme make the interior of this home feel cohesive, and smartly placed hooks ensure there is a space for everything, like the cutting board hanging on the wall.

LEFT: The Roost18 is one of Perch & Nest's smallest tiny home offerings, at just 18' (5.5m) long.

BOTTOM LEFT: Even in a small space, this home has the luxury of a deep bathtub.

BELOW: There is cushioned bench seating by the front windows and extra storage space in the eaves above the porch.

Courtesy of Perch & Nest (www.perchandnest.com) and Tiny Planet Design.

Courtesy of Perch & Nest (www.perchandnest.com) and Tiny Planet Design.

Courtesy of Perch & Nest (www.perchandnest.com) and Tiny Planet Design.

Courtesy of Perch & Nest (www.perchandnest.com) and Tiny Planet Design.

TOP LEFT: A comfortable staircase—more convenient than a ladder!—provides access to the loft sleeping area.

TOP RIGHT: The loft sleeps two and includes a skylight to keep the space from feeling claustrophobic.

BOTTOM LEFT: Unbroken vertical space above the kitchen makes meal prep a breeze and keeps the home from feeling cramped.

BOTTOM RIGHT: A stylish storage nook next to the stove allows easy access to the minimal utensils needed to cook.

RIGHT: The sleek black exterior of this 26' (8m) home, the Roost26, belies its bright and spacious interior.

BOTTOM LEFT: An incinerating toilet doesn't look out of place in this modern bathroom.

BOTTOM RIGHT: A subtle induction stovetop ensures you can prepare your meals quickly and easily.

Courtesy of Perch & Nest (www.perchandnest.com) and Tiny Planet Design.

Courtesy of Perch & Nest (www.perchandnest.com) and Tiny Planet Design.

Courtesy of Perch & Nest (www.perchandnest.com) and Tiny Planet Design.

TOP LEFT: There's no need to miss your favorite show—the television seamlessly fits onto the wall, where it won't take space away from the sleeping area.

BOTTOM LEFT: You won't encounter any of the annoying ventilation problems that often plague tiny homes with the open-air layout above the rooms in this model.

TOP RIGHT: You'll forget you're in a small space because of all the light from the windows and the white finishings.

BOTTOM RIGHT: Giant front doors to a shaded porch ensure you can enjoy whatever weather the world throws at you.

RIGHT: Relatively giant for a tiny house, the Roost36 model is 36' (11m) long, but it is still completely mobile thanks to its triple axles.

BELOW: View from the kitchen toward the front of the house. A full-length sofa allows total relaxation. Built-in shelves ensure that even though a large chunk of the wall is taken up by windows, there is still plenty of space for all your items.

TOP: The view from the front door reveals almost the entire home at a glance. The relatively large size of this home means a stylish and large fridge can live here, too.

MIDDLE: With plenty of headroom and space along the sides, there is no claustrophobia in this classic loft.

BOTTOM LEFT: This front deck allows you to enjoy your own personal "yard," rain or shine.

BOTTOM RIGHT: This genius bathroom includes a handy tray that acts as movable storage whenever you are using any of the facilities.

THE JOY OF TINY HOUSE LIVING

Courtesy of Perch & Nest (www.perchandnest.com) and Tiny Planet Design.

Courtesy of Perch & Nest (www.perchandnest.com) and Tiny Planet Design.

Courtesy of Perch & Nest (www.perchandnest.com) and Tiny Planet Design.

TOP: The 20' (6m) Pecan is a great example of a small space done right.

BOTTOM LEFT: There's a nice view looking down from the sleeping loft. You can see how the natural light enters from all sides of this home and brightens the space.

BOTTOM RIGHT: For such a small home, the loft is still quite large, sleeping two with minimal worries about bumping your head.

TOP: The view from the front door reveals most of this home, including a space-saving ladder to the sleeping loft and an inconspicuous bathroom in the back.

BOTTOM LEFT: A cozy seating nook tucked into the front of the home means there's always a place to relax for a little while.

BOTTOM RIGHT: The kitchen is quite minimal in this small home, with no stovetop. You can still manage nicely with what is there.

TOP: Jenna Spesard of Tiny House Giant Journey downsized her life and built her 165-square-foot (15-square-meter) home in about a year in order to travel around the country and the world.

BOTTOM LEFT: There's room for a woodstove for heating and cooking, a furry friend, and a subtle staircase that also serves as pantry and closet.

BOTTOM RIGHT: A well-lit loft is the perfect place to rest your head after a day of travel or work. As a bonus, there is roof access from the loft—on a nice day, the roof is a great place to relax.

TOP LEFT: Even part of the ceiling is used ingeniously as storage space for snowboards. Jenna spends most of her work time in the seating area at the front, where USB charging ports are plentiful.

TOP RIGHT: Corrugated metal sheet makes a surprising but stylish finish for the bathroom, which includes a composting toilet and a small tub that Jenna uses to give her dog a bath.

BOTTOM LEFT: The distressed reclaimed wood details in the interior make the home feel like a cozy bungalow.

CHAPTER 1

FUNDAMENTALS

Before you run out and get yourself a tiny house, there are a few things that you should ponder. It's potentially a big, life-changing step, and being a bit introspective now will pay dividends down the road. People wind up gravitating toward tiny houses for different reasons. You should ask yourself, and be clear about, what your motivations are. This chapter will help you sort through some of those drivers.

HOW I ENDED UP IN A TINY HOUSE

I want to give you a little bit of background on who I am and why I'm writing this book. I'm not doing this because you should be particularly interested in my personal story. My story, however, is probably not unique—chances are you'll see some parallels with your own life.

I grew up in Canada, and my earliest memories are of being outdoors fishing, camping, hiking, and doing a lot of fun stuff. As an only child, I was lucky to have two parents who focused on me, and I got to spend a lot of quality time with them. In my teenage years, I wound up in the New York City area, and then,

My daughter running around outside of one of my gypsy wagon builds up in the Catskills.

after moving around some more, including a stint in Europe, I found myself gravitating toward the New York City area again. Subsequently, I got married and had a child of my own. It was at that point when I realized the environment that I was living in was very much a suburban, densely populated, and congested area, and that this was very different from my experience as a kid. There were simply not enough trees.

As a father, I wanted my daughter to have some of the same experiences that I had had when I was younger. I started to think about purchasing a piece of land somewhere reasonably close by where we could spend some time. The idea was to camp and have a place to escape to from the nuttiness of the outskirts of New York City. I found an area not too far away in the southern region of New York known as the Catskills. We could easily get there for an overnight trip or even a day trip. Once I had purchased a property there, though, I wasn't sure what to do next. It was a great location, it was on parkland, and there were about three and a half acres to roam and explore. There were trees galore—I now owned the woods. Well, a tiny little part of it, at least.

Then I discovered Jay Shafer's *The Small House Book*. I don't remember exactly how it fell into my hands, but sometimes when we look back at things, we realize that they were just meant to be. I started looking through that book and was very intrigued. I wasn't a huge fan of tiny houses on wheels at that point, but I did like the concept of a cabin or a small building built on a foundation. That book got me thinking about the land that I had purchased. One nuance of the land was that it was zoned for recreational use only, which meant that you couldn't build a substantial, permanent structure on it. You could put up a shed, or a pagoda of sorts, but you could not build an actual house, even a miniature one. This recreational zoning of the land made me revisit my views on tiny houses on wheels, because tiny houses on wheels can get around a lot of zoning regulations by technically being a custom recreational vehicle (RV). They don't fall under the same restrictions as traditional building structures

Aerial drone shot over my Catskills, NY, property. It's beautiful up there.

My daughter lounging and reading in one of my tiny builds.

do, and they generally don't require planning/building approval.

With grand visions of a tiny house retreat in my head, I went so far as to purchase plans for a house from the Tumbleweed Tiny House Company, which at the time was the premier supplier of all things tiny house related. I got the plans, I looked them over carefully, and I realized it was a somewhat daunting task to build a tiny house on wheels. I don't come from a construction background, and these were architectural plans that had many symbols and details on them that were not necessarily easy for me to decipher.

Right around this time, Tumbleweed started offering something called an "Amish Barn Raiser." They were effectively selling you the trailer and doing all the framing, sheathing, and whatever else you wanted them to do on top of that trailer, short of providing a totally finished house. This was fantastic, because not only did it eliminate some of the fear that I had about doing the framing correctly, but it was also going to save me hundreds of hours of labor and travel back and forth between New Jersey and New York.

Hauling my Tumbleweed tiny house shell back to the East Coast from Colorado. At this point I'm somewhere in Kansas.

I didn't hesitate to place my order. The build took place, very inconveniently for me, in Colorado, about two thousand miles away from where I was. They did offer a delivery option, but I chose to fly out there and rent a U-Haul truck to tow the house back to the East Coast. All and all, this was a pretty bad idea, because towing a house was a lot for the truck to handle, and the distance and the speed limitations made for a very long and arduous four-day journey. I did eventually get the house back to the East Coast in one piece, where I proceeded to work on it.

Throughout this book, I'll relate some of my experiences regarding the aspects of building a tiny house, so I'll leave those details for later, but I did eventually complete my first build. In fact, I finished it just in time for a tiny house show in New Jersey that was being hosted and presented by the United Tiny House Association. In the span of the several years I spent working on the house, perhaps only

Exhaustion and elation sometimes go hand in hand. I was thrilled and somewhat overwhelmed to have won this award.

People lining up to tour my tiny house at the United Tiny House Show in New Jersey in 2017.

There was a constant flow of people. I lost my voice at some point. It was a tremendous, life-changing weekend for me.

about ten people got to see the work in progress. At the show, however, around three to four thousand people passed through my house during the course of the three-day event. The whole show was pretty overwhelming, and the feedback that folks were giving me was very humbling and something that I hadn't expected.

My house won best tiny house at the show. Winning was something that I would never have expected. After the show, I took the tiny house back to my property and returned to work, but, in the back of my head, I had realized that this was something I truly enjoyed, more than my dreary nine-to-five corporate existence and massive commute. I was working in advertising technology, which was certainly not a field that I was passionate about. I had many roles in various capacities, but it was always the people that I liked, not the work itself. Sure, I was well paid, but, as many of you know, money

does not always lead to satisfaction—in fact, it rarely does. I started working on getting my life-coaching credentials, which is something that I saw myself perhaps doing on the side to help people.

Everything came together. I didn't really like my job; I had seen that the construction of my tiny house had garnered so much great feedback; and my start down the path of becoming a life coach had also shown me that perhaps there was a life outside of the technology field that I had found myself in.

Here's the key: At some point in all our lives, there's a chance to take a leap of faith, to believe in ourselves and do something that doesn't necessarily feel comfortable (like writing a book). That's what I did. I quit my job and decided that I was going to focus on life coaching and tiny house construction. The house that I built became a fixture in my life and has taken on a much more prominent role than I could ever have expected.

It's been about two years now since I left the corporate world, and I haven't regretted any of it. Sure, it's a little bit more difficult, and it's sometimes tricky to structure your days around work and to be disciplined to do what needs doing on a daily basis. But my leap of faith has worked out rather well for me. I'm writing this book to share part of my story and my experience around tiny houses, and I encourage anyone and everyone to consider taking their own leaps of faith, whatever that may look like.

We're only on this planet for a certain amount of time, and it's the things that we don't do that we generally regret, rather than the things that we attempt. Believe in yourself; you can do it, and you have so many resources that can help you do it: YouTube, various websites, and what has become a very tight-knit tiny house community are all great assets for your tiny house journey. Ask yourself, what do you want from life? What is your goal? If you find yourself stuck in a role or a job that you don't like, then for your sake, please do something about it.

Anyway, that's a bit on my background and what got me started in this field. Like I said at the onset, your story may be quite different from mine, but hopefully you can take away a piece of my experience.

The interior of my first tiny house. I kept everything to a fairly cohesive theme. I'm still very happy with how it turned out.

A TINY HISTORY OF TINY HOUSES

You may think that people have discovered tiny houses somewhat recently. You've perhaps seen a TV show or pictures of tiny houses online, and this is what piqued your interest. The reality, though, is that tiny homes and small living have been around for a very, very long time. The current tiny house trend is more of a rediscovery of living little rather than a purely new phenomenon.

Human beings have always lived tiny; throughout **most of history**, people have had rather small dwellings. You can go all the way back to caves and teepees, or consider a time in the United States, when people were traveling across the country for months in covered wagons. If you look at older cities in North America, like Montreal, you'll see various older neighborhoods all featuring compact houses, and the same is definitely true in Europe, which is packed with neighborhoods like these. Small houses were built because it was difficult and costly to obtain materials. Current notions of disposable income didn't exist; the average person did not live in a world where it was feasible to build larger and larger houses just for the pleasure of it.

The original tiny house on wheels: a covered wagon. Well-known for their role in transporting settlers along the Oregon Trail in the 1800s, covered wagons were glorified storage trailers and not nearly as comfortable as the tiny house you will build.

Tiny houses like these in Lüneburg, Germany, were built for centuries throughout Europe.

Some houses in Iceland, like these from the 1800s, were integrated into the land precisely because there were minimal building materials available for construction. Due to the island's harsh climate, there are no trees that are suitable for constructing houses. The only wood that used to be available was the kind that washed up on the shores as driftwood.

In the United States in the mid-20th century, though, **disposable incomes** were more common and more considerable, and it became feasible for the emerging middle class to live in larger and larger houses. Many people started wanting bigger, more lavish homes to show off their success (and sometimes show up their neighbors). This culminated in the McMansion phase that we saw before the housing bubble burst and the recession of 2008. At that time, people rediscovered the tiny living phenomenon, often because they had a certain amount of insecurity with their finances. People also realized that the trend toward ever-bigger houses was **not truly serving them**. Ever-larger homes benefit towns because of the higher taxes that towns can charge, and the construction companies that build larger houses also derive greater profit from these more substantial structures. There is also a knock-on effect in furniture: since larger homes need more furniture,

furniture companies benefit. Larger houses also require additional heating and cooling, so energy providers see a benefit as well. There is a lot of interest from various outside entities to encourage people to live in bigger houses. But in the end, there is little in the interest of the actual people living in them.

When we talk about the modern tiny house movement, one person stands out: **Jay Shafer**. In the late 90s, Shafer, for a variety of reasons, decided that he wanted to downsize his life and live in a smaller sort of way. He was able to do that by creating a very nicely designed tiny house. He documented his process and put it out to the world, and it started to resonate. People could clearly see the cost-effectiveness and efficiency in going tiny, and it appealed to a lot of different people for different reasons. **Dee Williams** is another one of the originators of the tiny house movement. She downsized her life into a minimal amount

The originator of the modern tiny house movement: Jay Shafer (right). It was an honor to meet the man who has inspired so many others to downsize their lives.

of space. These are the pioneers of the new tiny house movement.

Now we are in the midst of the modern tiny house movement, and it's picking up traction. Since it's such a significant shift away from the prior notions of building ever-larger houses, it has been somewhat difficult for certain aspects of the movement to keep up with the times—or, perhaps more accurately, it has been difficult for the times to keep up with the movement. The most noticeable aspect in this regard is that the **legalities of living tiny** are not in line with the existing models of government and structures that are in place. For example, many towns and municipalities have minimum size restrictions on living spaces. This means that for you to be able to build a house, it has to have a certain number of square feet for it to be recognized as a house. These kinds of policies were put into place for a variety of reasons, some of which have to do with safety, some of which have to do with taxation, and some of which have to do with towns wanting to have uniformity in the look and feel of their housing.

We'll go into the current state of tiny house legality later, in chapter 5 (page 144). For now, though, I hope you can see how the tiny house movement isn't just a newfangled fad—it's actually a logical reaction to how housing has been changing over the years, and a callback to different eras with different standards. The modern motivations for going tiny are as varied as the people who do it; read on to learn more.

With the right amount of activism and pressure, the world will make room for tiny houses. Thanks to people within the movement, like Alexis Stephens and Christian Parsons, co-founders of Tiny House Expedition (whose house is pictured at right), progress continues to be made.

WHY ARE PEOPLE GOING TINY?

Tiny represents different things to different people at different points in their life. Here I'll define what it tends to mean for different age brackets and demographics.

Tiny houses have a specific appeal for **younger people**, people who are coming out of college or maybe just starting off down a career path. Going tiny represents an affordable accommodation during a period when financial stability might not yet be established or possible. Younger adults coming out of college may be in debt from paying for their education. Taking on additional debt in the form of a 30-year mortgage does not appeal to these young adults. This younger generation is gravitating toward smaller, more affordable housing, given that the large urban centers in the United States and elsewhere are experiencing affordable housing crises.

Tiny houses (and other alternative structures such as buses, yurts, or vans) therefore become a natural and clear alternative to traditional living for young people. When you combine the affordability with the ability to easily relocate a tiny house—to move it wherever the work is—it is an appealing proposition. More and more professions and jobs can be executed remotely, and this is another reason why living in a tiny house is a good alternative. According to a Gallup survey, more than 43 percent of workers did some of their work remotely in 2018. This trend keeps going

This small, towable home would appeal to many members of the younger generation, who are often motivated to go tiny because the cost is considerably cheaper than a 30-year mortgage.

Tiny houses are an opportunity to rethink what it means to have a home.

up year after year. That tiny house can be situated virtually anywhere, as long as the person has access to the Internet and to power and therefore can perform any number of different job functions.

For people who are **middle-aged**, tiny living represents many of the same benefits. But there are also certain life situations in middle age that make living tiny more viable. People in the middle of their lives often go through major life transitions, like the dissolution of a marriage (according to Pew Research Center, divorce rates consistently hover just below 50 percent), or a new career with different needs. There are many reasons why someone might be looking to downsize and save money in their thirties and forties. Plus, the same benefits of mobility and remote work that apply to the younger generation certainly hold true for middle-aged folks as well.

When you get to the **older generations**, many people may be looking to downsize or travel more. Their children may have moved out of the house, and now they're living either by themselves or with a partner in a house that has outgrown their needs. Downsizing, the empty nest phenomenon, saving money, and being able to relocate are main drivers for folks in their fifties, sixties, and even seventies.

Cutting across all of these generational aspects is the element of people **rethinking what it means to have a house**. The standard notion of traditional housing doesn't really appeal to everyone anymore. You may be living in a space not really designed for your personal needs. The tiny house movement is founded on creating a personalized space and building only what you need. When you build according to your vision, and not someone else's, you invariably create something that is uniquely your own. Homes have always been an expression of who we are; we decorate them a certain way, we fill them with certain furniture, and we paint them certain colors. Tiny houses take it a step further: not only can

Why Are People Going Tiny?

you personalize the decoration and the fixtures, but also the entire design. It's your vision, not someone else's, and this element of customization is a really appealing notion for people of any age.

One of the other benefits—and some people's main goal—of going tiny is that you are **lowering your carbon footprint**. We all have an impact on this planet, and being able to reduce that in some fashion can be appealing. Another great by-product of the tiny movement is that people who have gone tiny tend to have more **disposable income**. For most tiny house owners, this disposable income will not be spent on acquiring more consumer goods, but rather on gaining new experiences. (I cover this in more detail on page 42.)

One reason that anyone, at any age, may be motivated to build a tiny house is that, unlike in traditional housing, every aspect of it is customizable. For example, I made these unique mock shutters to decorate the windows on one of my houses.

Solar panels and other forms of renewable energy can help reduce your carbon footprint in a tiny house.

IS TINY RIGHT FOR YOU?

You have to ask yourself many questions when you are considering going tiny. The main one that I want you to answer for yourself is, "Why am I doing this in the first place?" If you don't have a good answer to this question, you need to find one, because it is important to have a clear answer before you start down this path. Beyond this essential question, there are a host of other somewhat philosophical, somewhat practical questions you should ask yourself as you consider if going tiny is right for you. Feel free to use the space at the end of this book to jot down your answers.

- What appeals to you about tiny houses?

- What's your life situation? Is it part of the reason why you want to get a tiny house? Perhaps you're an empty nester; perhaps you've gone through a divorce; perhaps you're looking to downsize for some reason; or maybe you're retiring.

- How will a tiny house make your life better? I know that seems like a strange question, but if a tiny house isn't going to make your life better in some way, then why do it?

- What do you hope to achieve with a tiny house?

- What about the timing? What makes you want to do it now?

- Do you have concerns about going tiny? What are they? How will you address them? There may be a chapter in this book to help you out.

- Is it just you who wants to go tiny? Are friends and family supportive or unsupportive?

- Who else is impacted by your desire to go tiny—a significant other, children, family, friends? Are these people going to go down this path with you or alongside you, or what role will they play in the process?

- What's holding you back? This is a huge one, because many people want to live in a tiny house, but they never take that initial first step down the path to owning one. Is it purely financial? Is it something more vague, like maybe this is not the right time, or any other excuse we so quickly come up with for ourselves?

Knowing your ideals will give you solid direction. For example, if you want this to be your morning view—if exploration and travel are the ideals you are pursuing—then you'll want to make design decisions that allow you to get on the road quickly and easily.

Once you've established your motivations and goals for tiny house living, there are also practicalities to consider in relation to building a tiny house.

- Do you feel like you're able to build it yourself, or will you be looking for someone to build it for you? In other words, are you going to outsource the process? If you are going to outsource the building and the timeline, there's a high likelihood that this transition is going to cost you quite a bit more than it would if you were doing it yourself. You'll find more detail about hiring a contractor on page 83.

- What do your finances look like? Are you selling a traditional house or something else to fund going tiny? Is it a prerequisite that you sell something before you can start? Will you be taking out a loan? There are more and more builders that offer financing for the tiny houses that they build. Perhaps you have the cash available to buy the tiny house outright, which is the best-case scenario. Regardless, you need to think about the financial implications of living tiny. You'll find more detail about financing on page 78.

- Don't forget—getting the tiny house is only half the battle, because you also have to park it somewhere. What are the local regulations for tiny houses in your area (or the area you want to move to)? It may seem like tons of people are building tiny houses and living in them, and that is definitely happening, but a lot of it is happening in a legal gray area where people are living under the radar, perhaps parked in someone's backyard (read: legally owned property) somewhere. As tiny houses are garnering more and more interest, they're becoming legal in more and more places—or the gray areas are disappearing. The path of least resistance is often to get legal clearance for tiny houses as Accessory Dwelling Units, or ADUs. This allows a tiny house to be placed on the property where another house resides, acting as an extension and expansion of living space. ADUs are legal in an ever-increasing number of places, such as Los Angeles, parts of Colorado, and New Hampshire. Look into your local laws by contacting housing and zoning commissions. You'll find more detail about legalities in Chapter 5 (page 144).

This tiny home sells for $48,000. Can you afford to buy a home like this outright, or with a loan? Or would your goals be better served by building a house yourself?

You may be utterly fascinated by the cute, modern look of tiny houses—but don't make the mistake of assuming that means they are a good fit for you. Do you see any closets, or a washing machine or dishwasher, or a bathtub, in this stylish but tiny interior? Can you make the necessary sacrifices to live tiny?

Along with the plan for how you're going to finance and where you're going to put your tiny house, you'll need to consider the material aspects of your endeavor. Tiny house living is an expression of what's important to you, and when you only have a small space, you can't have everything. This is a vital question: **What is essential for you in a home?** You will have to make tradeoffs; you will have to think very carefully about what the house has to have. If you can't live without a dishwasher, then your tiny house should have a dishwasher; but you must realize that for every item you consider essential, there's going to be a cost, either financially or spatially or both. There will be things that you will initially feel like you need, but think about whether they are even feasible in a tiny house. If you truly need four bedrooms or a huge walk-in closet to hold your hundreds of articles of clothing, then tiny living is probably not for you. This brings us to another vital question: **Are you able to part with most of your stuff?** However you're currently living, it most likely affords you the ability to own many more possessions than you can if you live in a tiny house.

The final question to consider when deciding whether tiny is right for you brings us back full circle to the beginning. Imagine yourself in your tiny house. **What do you look forward to the most about living tiny?** Think about this and take a minute to jot down your answers. Is it the ability to move your house to another location if you choose to? Is it the fact that you're going to feel more comfortable in a smaller space with fewer things and fewer burdens? Is it the financial freedom that many people enjoy in the tiny house community? Whatever the case may be, whatever your personal answer is, make sure that you have a good understanding of where you are and how you fit into the tiny house equation. "Well, I like tiny houses, and that's why I want one," is not a good enough answer.

I hope all the questions we've raised here have given you plenty of food for thought and some things to work out for yourself. Getting them straight in your head will allow you to plan for your tiny house effectively.

EXPERIENCES VERSUS STUFF: MINIMALISM

The American Dream has always emphasized going big and doing better: go to college, buy a car, get married, buy a house, have kids, retire, and so on. There are certain elements of this philosophy that we've all adopted; our parents likely bought into this idea as well. But in the modern day, some people are concluding that this exact vision is not necessarily what they want out of life.

Part of the American Dream in the last century has been focused on consumerism and consumption. As a result, most of us have acquired way too much stuff for our own good. We have houses that are too big and filled with too many things. We have garages that aren't used for parking cars but for housing our overflow of stuff. And when there simply isn't enough space in the house, well, off-site self-storage is a booming industry in the United States. There are over 50,000 self-storage facilities in the U.S. that bring in over $30 billion in annual revenue. Many people end up living to work instead of working to live because they have big mortgages and other financial commitments. After acquiring more and more possessions and tying themselves down financially, many people have started to wonder whether this is indeed what they want to be doing. They ask, "How do I break this cycle? How do I lower my bills? How do I decouple from this never-ending rat race? Is it making me happy?"

Going tiny is a process of self-discovery; it's about figuring out what's fulfilling in life to you, other than possessions. I'm sure you've already heard of the notion of experiences being more valuable than stuff. We all have stuff, but when you spend your money and time on more experiences instead of more stuff, you enhance your life. This trading of things for experiences is quite rewarding. And the side effect of

Less is very often more.

prioritizing experiences over stuff, living clutter-free, for many people equates to living worry-free. There's been a great deal of interest recently in books that help you downsize your possessions. Letting go of things and assessing what's truly important to you is a process that you have to take on personally, figuring out where you fit into the minimalism spectrum.

As we discussed in the previous section, tiny houses are a lot smaller, and therefore the things that you put into your tiny house have to, by their very nature, be important to you. Otherwise, there's simply no room for them. As you go through this process of deciding what experiences you value, what it is you're trading your possessions for, you will learn what's important to you, find out who you are, define what you want from your life, and take steps toward that.

Living Clutter Free: Downsizing

How do you start down the path of living without clutter? Based on my experience, I advise you to do it slowly and take it one step at a time. One good way to start is to take just ten minutes each day to look around the space you're currently occupying and find things that you don't need. Sometimes these things are tucked away in a closet; sometimes they are shoved under a bed. Think about things that you haven't used in a while (for weeks, months, or even years). We all have those items. If you haven't used (or seen) something in months, do you need it? Seek out these unnecessary items and dispose of them, either by recycling them, selling them, donating them, or throwing them away.

Here are a few focus areas you can tackle in your quest to downsize.

Kitchen: Are there utensils and appliances that you don't use anymore? Kitchenware is infamous for one-use tools; if it only serves one specific purpose, can you do without it? How do you, personally, cook? Think about how you prepare meals—and I'm not talking about meals that you make once a year on a special occasion. I'm talking about your everyday life. How many cups and mugs and plates and spoons and bowls and pots and spatulas do you have? Count them; the number may shock you. Do you need them all?

Wardrobe: Clothing is another excellent place to purge. We all have those items that don't fit us anymore, or that have gone out of style, but we still like holding onto them; they continue to hang in our closet year after year, untouched. Some of us have too many shoes, and we likely don't need or wear all of them. We also have highly seasonal clothing only suitable for certain times of the year, instead of pieces that are versatile. Considering all of this, what can you get rid of?

Media: Media takes up a ton of room in many people's homes. If you have an extensive CD or DVD collection, could you switch to virtual copies or subscription services instead? This potential for space savings also holds true for books. I know many people like to have a book collection, and I'm no different in that regard; I certainly have more books than I need. More and more, though, I find myself either reading a book and then passing it on to someone else, or reading an electronic version of a book. Give it a try.

We all have something like this clutter at home—and for some of us, it's our entire home. Are you prepared to cut down on your worldly possessions to live tiny? Are you prepared to prioritize only what matters most to you?

You may be proud of your music collection, and rightfully so. But consider going digital—media is one of the easiest categories in which you can reduce your possessions in order to fit into a tiny house.

Experiences versus Stuff: Minimalism

INTERVIEW WITH

ALEX EAVES

Alex: I am Alex Eaves, and I am the Reuse brand owner, filmmaker, and 98-square-foot (9-square-meter) tiny house dweller living inside a used box truck.

Chris: *What made you choose a box truck versus any of the many other options that tiny living enthusiasts have available to them?*

Alex: I had come back from a concert tour in my Pontiac Vibe. I needed an upgrade; I was traveling in the Vibe around the country and sleeping in it sometimes. Derek "Deek" Diedricksen suggested the box truck, and as soon as he said it, I was like, "Oh man, that's perfect," because when I used to travel with vans selling merchandise, we would rent box trucks and sometimes sleep in them. We'd put a mattress or a big stack of T-shirts and sweatshirts on the floor and then sleep on it, so it was full-circle perfect. Deek went on to be instrumental as the co-creator and builder on this joint project.

Alex Eaves (right), the owner of the Reuse box truck tiny house.

Chris: *How long have you been living in this rather luxurious interior here in your box truck?*

Alex: It's almost a year.

Chris: *What's the most surprising thing that you didn't anticipate that you found out during this year?*

Alex: That the number one adjective people use when they come in here is "spacious."

Chris: *Reuse is important to you. For your box truck, you used mostly, if not all, repurposed materials, correct?*

Alex: Yes, everything in here is pre-owned, used, upcycled, or repurposed, except for the caulking, the spray foam in the corners and around that glass door, the rubberized paint in the bathroom, and a few specialty screws. I created a documentary on the subject of reuse, which you can find at *www.reusedocumentary.com*. We're also in the process of making a documentary about the truck.

Front view of Alex's box truck conversion.

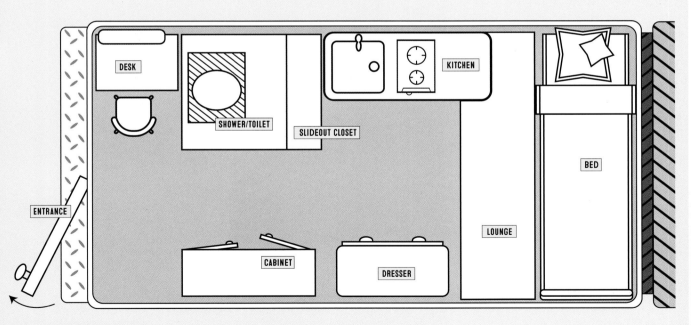

Alex Eaves' box truck floor plan.

Chris: *There are clear advantages to using repurposed materials that immediately come to mind. But what's the downside, if any, of using repurposed materials?*

Alex: Acquiring some specific items can be difficult. For the most part, for this box truck, it worked out just fine, but just a couple things, like the ⅜" (1cm)–thick plywood that is on the ceiling, were tough, since we ran out. And sometimes there are too many options. When people heard what we were doing, they were like, "Come to the house, check out this antique wood I have." All cool little things that could have been in here that we turned down.

Chris: *In a nutshell, it can slow the build process down if you're relying on finding, sourcing, and acquiring some of these materials.*

Alex: Yeah, but I think that's what makes it fun, too. If you want it done ASAP, then reuse may not be for you. If you're in it for the adventure and making this your own custom house, then it's totally for you. And we spent next to nothing. Almost everything you're looking at was free.

Rear view of Alex's box truck conversion.

Chris: *In terms of choosing a box truck, is there anything you would have done differently? A bigger box truck, a smaller box truck?*

Alex: I do get jealous when I see the bigger box trucks on the road. Overall, this is the perfect one for me, though, because it fits in a regular parking space. It makes getting around a lot easier.

Desk working area in the back of the Reuse box truck. That's a nice, big window.

Chris: *Everyone always wants to know where you can park something like this. What's your advice for finding a suitable spot, whether it's for a tiny house or a box truck or whatever it happens to be?*

Alex: Barter is a good option. I've got an agreement with my sister. She's currently redoing a house in Massachusetts, and that's where I park for the majority of the time. I'm kind of watching the house while they're living elsewhere. And then also when I'm on the road, I'll cook you dinner one night in exchange for a place to park. Box truck chili.

Repurposed doesn't mean uncomfortable. This nice interior doesn't feel small or cramped.

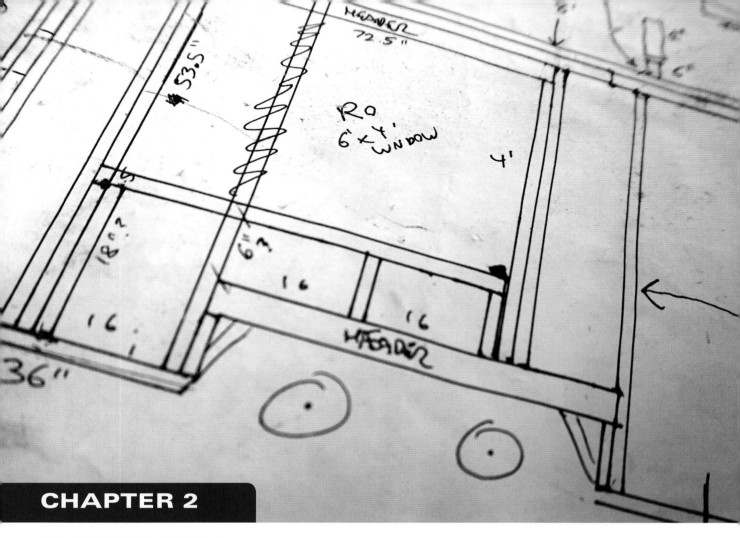

DESIGN

It's time to do some planning. You have sorted through all your reasoning for going tiny and have embraced the concept of downsizing your life. Now starts the fun part of figuring out what tiny living will look like for you: how your tiny house should be structured and what it should have. I'll guide you through some of the things you need to decide before moving further. Having clarity about these elements will make the construction phase easier.

WHAT'S IMPORTANT TO YOU? START WITH A VISION

Defining what's important to you is essential for happy tiny house life. You have to think about all the various elements that your tiny house needs; invariably it will have most or all of the "rooms" that a traditional house would have (kitchen, bathroom, bedroom, and living room). You need to **start with a vision** so that you have direction when it comes time to design or choose a design for your house.

Think about what kind of bedroom you like. Think about what kind of kitchen you like. Think about things in the bathroom that are important to you. Do you need a tub or are you okay with a shower? Are

Yes, you can have a tub, if that's what you really want and need in your tiny house.

A minimal kitchen I installed in a recent build. This house is intended as an AirBnB. This kitchen will fully serve that purpose.

Style never needs to be compromised. An amazing bathroom in a house by Craft & Sprout.

I searched far and wide for the ideal "kitchen." This piece of furniture blended in well with the rest of the house's interior.

you into big, oversized, comfortable furniture? Well, that might not work so well in a tiny house. How will you tackle laundry? A washing machine and dryer tend to take up too much space—do you really need them? What kind of toilet do you want to have? Are you okay with a composting toilet, or do you want a traditional toilet? How are you going to heat your tiny house? Will it be propane or electric? What is the environment in which you live?

At this point you're not getting into the practicalities, per se, but rather envisioning what it is that you want. (Specifics for all of these considerations are discussed elsewhere in this book.) What you need to realize is that in a tiny house, everything that you want and need is generally accomplished with some form of **tradeoff**. If you're going to take up space in the tiny house with something that's important to you, you're going to have to minimize the importance of something else. An excellent place to start is to take everything out of the equation that you don't want or don't need. Alternatively, make a list of the things that you can't live without. This is your one chance to create a space for yourself that is uniquely for you. Thinking about it in these terms helps you get to the final decisions about what is important to you and what isn't.

Other elements to think about include **sustainability** and how **connected** to the world you want to be. Should this house be independent of external power sources and external water sources? You can have solar panels on the roof to supply the bulk of your electricity, for example, but understand that heating and cooling a house with solar isn't necessarily a viable or efficient method in all places.

Also understand that **size and mobility** are intricately linked. The larger you build the house, the less mobile it's going to be. Sometimes people lose track of these things when they're making their decisions and opt for aspects within a tiny house without considering the final impact. If you think of things in a weight-per-square-foot kind of way, it will help you make better weight-saving choices. For example, when outfitting your bathroom, remember that tile is heavy; sheets of aluminum are not. That is one of the reasons you commonly see metal-lined showers in tiny houses. When it comes to flooring, it is usually not designed to be light. Thick, solid wood floors will weigh the most. Engineered wood flooring is still actual wood (not laminate), but it is much thinner, which means lighter. In general, be wary of weight. It may not seem to matter a little here and a little there, but the cumulative effect can be significant. In my tiny house, I opted to use a solid wood sideboard buffet, the type that would typically sit in someone's dining room, and turn that rather substantial piece of furniture into my tiny home kitchen. There's space for a dorm room fridge in one of the cabinets, and I did cutouts for both the sink and for the cooktop. I chose this particular option because my tiny house was never intended to move very often; travel was not a priority, so I could afford a heavier piece of furniture. But it's these kinds of details that I suggest you spend time thinking about until you define, very intricately and very specifically, a personal vision for your tiny home.

Craft & Sprout bring amazing details to their spaces. There is no feeling of compromise here.

ROUGHING OUT A PLAN

As you refine your vision for your tiny house, you can begin roughing out a plan. You can start to think about whether to **build it yourself** versus **hiring a builder** (more details on this on pages 80 and 83). If you are providing your own labor, construction will cost half as much as having someone else build it for you. Labor, expertise, and time are all expensive. If you're going to build the house yourself, you're going to need to acquire some of the expertise to do the job, unless you're already totally qualified in construction.

Also important at this point is to set a **timeline** for yourself. It's easy to get sidetracked; life gets in the way, and it's human nature to put things off if we don't place them on some form of a timeline. Sit down and decide in what time frame you're looking to complete your tiny house.

The time you will need to complete a tiny house varies greatly based on the scope of the house, how much time you can dedicate to the task, and what your level of building experience is. Other factors can include sourcing of materials and even weather conditions if you don't have the luxury of building a tiny house indoors. Some of the smaller houses that my company builds—like a 50-square-foot (4.5-square-meter) custom camper—take me around four weeks to complete while working full time on them. Some of the larger houses I build take a few months—like a 16' (4.8m) long single-level build. The original tiny house I built for my daughter and me took several years to build: the build site was 100 miles (160km) away from where I lived, I was working a full-time job and could only tackle the house on the weekends, and harsh winter weather prevented work for a few months out of the year. You can see how all these factors conspire to create vastly different time frames for different people and scenarios.

To get to this stage, you need to take care of the basics first.

Where your tiny home is going to be built is also an important factor here. If you're doing the building, you're going to need a space in which to build. It could be someone's driveway, or maybe you know someone who owns a piece of land somewhere. If you're in a more urban environment, look around and see if there is a parking lot or a storage yard behind some semi-industrial facility that will allow you to pursue your endeavor.

Finally, as you consider the building process, you should be willing to stray at least a little from your initial vision for your tiny house. As the Rolling Stones once said, "You can't always get what you want … but you get what you need." Keep an **open mind** and don't fixate too much on any one feature. There are various reasons for your plans to evolve away from your original vision. Accept this and understand that changes will and should happen. The two elements on which you should be less flexible are your timeline and your budget, because those have direct impacts on the process and the completion of the project—and quite possibly on you having a roof over your head!

In the end, whether you're building the house yourself or having the house built, educate yourself. Make sure you know what you're getting into, have considered all variables and costs, and can proceed with confidence.

CONSIDERING MOBILITY

Regardless of who is building your tiny house, there are two key components that you need to be very clear about at this stage: size and weight. I've touched on them briefly before, but now that you're in the design phase, you'll need to give them some serious thought.

Let's start with the **height and width** considerations. The size of the house broadly impacts the towability. In order to move your house easily on public roads, limit yourself to 8' 6" (2.6m) in width and 13' 6" (4.1m) in height. (Don't worry too much about length, since you have a max of 40' [12.2m] for that.) These dimensions are not dictated at the federal level, but rather set at the state level. Some states allow for slightly larger maximums on the roads. If you do a web search for "vehicle height restrictions" along with your state or geographic location, you will find specifics for where you live. Even if the limits happen to be higher in your area, though, I strongly advise you to stick with the limit I've stated here, since that will be the least restrictive. Exceeding the limit will

You can't take the subject of weight lightly. Your choice for each feature and fixture in your house, from porcelain sinks to metal gas stovetops and more, will impact the final weight of your construction and therefore how easy it is to tow it.

just make it harder to travel through certain states. Plus, going under a tight underpass while towing a tiny house is one of the scariest things you can do! If you do surpass these dimensions, you will need special permits to tow the structure down the highway, and there will be many roads that will be closed to you (such as those with low bridges and tunnels).

The materials that you use to build your house also impact towability due to **weight**. You could build a rather sizeable tiny house that is relatively light, and you could build a very small tiny house that is incredibly heavy. The ramifications of all this is that weight will determine what kind of tow vehicle you will need. If you are building a tiny house that's on the heavy side, you will need a hefty truck to tow it with. Here are some weight examples to give you some perspective. I build small camper/RV houses that weigh around 2,000 pounds (900kg). They are

One way to ensure mobility is to build your tiny home out of something that is already street legal, like a school bus.

small and, for a wood frame construction, fairly light for their size. This small gypsy wagon–style house can be pretty easily towed by an average SUV or small truck. Moving up from there, most of the houses you see pictured in this book are significantly heavier. On average, tiny houses fall somewhere in the 8,000 to 12,000 pound (3,600 to 5,400kg) range. That is for a 150 square foot (14 square meter), 20' to 24' (6 to 7.3m) long tiny house with a loft sleeping area. The biggest houses, in the 30'+ (9.1m+) realm, can weigh as much as 15,000 to 18,000 pounds (6,800 to 8,100kg). A massive diesel truck is needed to move those houses. You would likely need to spend as much to buy the truck to move that kind of house as to pay for the house!

You don't necessarily need to own a tow vehicle; depending on how often you intend to tow your tiny house, it may not make sense for you to have the ability to move it on your own. There are services you can hire to tow your tiny house from point A to point B, which generally charge on a per-mile basis.

If independence and mobility are key motivators of your going tiny in the first place, then you will want to build as small and light as possible. Some of the newer trucks on the market have fairly high towing capacities. Something like a newer Ford F150 is likely fine to tow the smaller end of the tiny house spectrum (5,000 to 8,000 pounds [2,200 to 3,600kg]). Anything heavier and you'll need something more like a Ford F250, which can handle upwards of 10,000 pounds (4,500kg). I'm using Ford as an example— all truck manufacturers have different models and classes. Inquire with your local dealer or the vehicle manufacturer about exact towing capacity. Whatever you do, stay within the towing range for your vehicle. Exceeding it is reckless and dangerous to you, your house, and the people you share the road with.

Realize, too, that when you are towing something, the tow vehicle's fuel economy will drop precipitously.

This will be felt at the gas pump and by your wallet. I go into the nuances of towing in a later chapter (page 137), but understand that mobility and size of house are tightly related.

If you want to avoid towing altogether, you could consider converting a **school bus** into a tiny house. Many have done this, and it's quite amazing how great a "Skoolie" conversion can become. (See page 140 for an interview with a man who lives in a school bus full-time.)

OFF-GRID LIVING

The notion of being completely disconnected from the rest of the world is appealing to some people, and going off-grid is becoming more and more popular and achievable. You can achieve isolation from not having nearby neighbors, but you can also seek isolation in the way that you get your energy and deal with waste. Traditional housing relies on having plumbing with municipal sewer connections to eliminate wastewater. It also means getting electricity from a large local electricity provider. There are ways

The sun and wind are well suited to provide a vast amount of power.

Off-Grid Living

to get around sourcing these essential services from local providers. The main three nontraditional energy sources are solar, wind, and hydropower. These are attractive because they are **sustainable**; all of your resources are being provided for by nature, which is great in terms of your carbon footprint and overall impact on the planet. We'll discuss these options in more detail in the section on electricity (page 95).

The **independence** that personally sourced sustainable energy brings is also quite appealing. You're no longer reliant on the power company to fix a downed power line after a storm, for example. The responsibility for the functioning of your energy sources lies with you.

Cost is another critical consideration when it comes to energy. When you don't have to pay large, industrial providers to deliver these resources to you, you'll enjoy significant savings over time. The initial setup costs of solar panels, wind systems, or hydro systems can be expensive, but over long-term usage, you will ultimately enjoy energy cost savings. At the time of writing this book, you can buy a 100W solar panel for around $150. When you add a charge controller and a modest battery, you can have a very basic solar setup for a few hundred dollars. That will likely only meet your most basic power needs, though. To have a tiny house solar setup that can power a few low-voltage appliances, venting, lighting, phone and laptop charging capability, etc., you are looking at spending more than $1,000 purely for the equipment. If you have someone install the hardware, your cost will double. Again, though, you're eliminating the need to pay an energy provider, and the cost of solar panels and batteries keeps coming down over time.

There are other ways to save energy and use energy efficiently in your tiny house that are determined by how and where you build the house. The better your house is built, and the better insulated it is, the fewer energy needs you will have, particularly related to heating and cooling. R-value is something that you will hear about; it is a representation of how well insulated your house is. Different insulating materials have different R-values, so pay attention to these when you are deciding how to insulate. Insulation is discussed in more detail in its own section (page 104).

There is also a notion of a **passive house**, a term that has emerged out of Europe where people are also looking to minimize their carbon footprint and impact on the environment. By designing a house so that it takes advantage of its natural environment, a more earth-friendly home is possible. For example, when the sun hits a house, if the house is designed in a passive way, it will absorb as much of the sun's power as possible. Especially in an environment where the weather is typically cold, the sun can then be relied on for heating. Conversely, if you're in a warmer environment, a passive house will be built and placed where it can reflect the sun's heat and radiation, helping to keep the interior of the house cool. The goal with a passive house is to adapt the structure to its given environment and create a situation where minimal, if any, artificial heating or cooling is needed to maintain a comfortable temperature.

ON WHEELS: THE FOUNDATION

The foundations of most tiny houses are on wheels. That means that you're building your house on a trailer. A question I often get from people is if the house can be removed from the trailer. The answer is generally going to be no, since proper and safe attachment of the house to the trailer is permanent. Whether you're building a traditional house or building a tiny house, if the foundation isn't strong, whatever you build on it is not going to be strong either. Read on to learn the ins and outs of trailer types and features to consider.

This single-axle trailer won't be able to handle as much weight as the double-axle trailer shown on page 58, but it will be more maneuverable.

Purpose-Built versus Adapted/Used

There are many places during a tiny house building project where you can save money. You can use repurposed materials, or you can source materials cheaply and effectively through the Internet. But regardless of how much you strive to save and cut costs on other aspects of your build, doing so with your trailer/foundation is not a good idea. Trailers can be purchased that are designed for tiny houses and are usually new; or they can be purchased used, in which case they are usually not designed for tiny houses. Is it possible to build a perfectly solid, good tiny house on a used trailer that wasn't made for tiny houses? Yes, it's possible, but it requires more work, knowledge, and skill than many first-time tiny house owners have.

Car hauler trailers may, on the surface, seem like a perfect foundation for a tiny house, and people do build tiny houses on them. Keep in mind, though, that since a car hauler trailer is not purpose-built for tiny houses, certain aspects are not ideal for tiny house construction. Although such trailers have a pressure-treated wood deck, they typically have a dropped slope at the back of the trailer that makes it easier to drive a car onto the trailer. There is also usually some form of ramp(s) that is welded on with large hinges on the back of the trailer. Both of these elements can make it harder to build a tiny house on top of the trailer. That drop-off doesn't lend itself very well to attaching framing to the tiny house, and the large hinges on the back can be difficult to remove because they have to be cut off by a saw or torch. Also, the exterior edges of a car hauler trailer may not be where you want them to be to attach the framing of a tiny house. Car hauler trailers may also not be capable of carrying the massive amount of weight of the tiny house you want to build. Trailers have one or more axles, and those axles are rated to a specific weight. Yes, cars are heavy, and there are car hauler trailers that can handle the weight of a tiny house, but not all of them can do the job.

Other trailers that you will come across are ones where the deck of the trailer is mounted above the wheels. **Deck-over trailers** raise the surface a considerable amount because the wheels and tires have to fit under the deck. This is not a good solution for tiny houses, because it limits how high you're able to build the interior space of the house. As I've mentioned previously, tiny houses are restricted to 13' 6" (4.1m) in height, and if you start out from a higher point above the wheels, you will have much less space and distance to work with until you hit that maximum allowable height.

The preferable solution to minimize this is a **deck between the wheels**. These are trailers where the tops of the wheels extend at least to the height of the surface of the deck and generally even a little bit higher. In other words, the platform that you will be building your house on is located between the wheels. The downside to this is that it introduces a part of the wheel well that will have some kind of design impact on the inside of your house. The framing may have to go around or over these wheel wells, and sometimes these sections lead to thinner walls that are harder to insulate and cover.

With all these considerations in mind, I recommend buying a purpose-built tiny house

A double-axle trailer I used for a recent build. This is a 16' (4.9m) trailer with a drop axle, so the building surface is very low to the ground.

trailer rather than repurposing a used trailer that was designed for a different purpose. Such trailers vary greatly in cost and quality. The bigger they get, though, the less the price increases, since it's just a little extra steel at that point. A good-quality tiny house trailer will start at around $3,000 for a smaller trailer and go up to around $7,000 for a larger one (for an 18' or 20' [5.5 or 6m] house). (These are rough numbers.) Also note that there are areas in the United States where these tiny house trailers tend to get built: the southeast, Ohio, and out west. If you live close to these areas, you're in luck. If not, you may have to plan a road trip or factor in shipping costs to get your trailer.

Trailer Components

Trailer **axles** are often either standard axles or drop axles. A standard axle is a straight bar with the wheels attached at either end. A drop axle is also a straight bar, but comes up on either side by a few inches before the wheel is attached. The wheel being mounted higher results in the trailer being lower. Therefore, drop axles will actually increase the amount of height that you can build onto the deck, since they lower the deck even further than a traditional trailer deck that's between the wheels. However, there is a tradeoff for everything you choose, and drop axles are no exception. Since they lower the house closer to the road surface, this can introduce problems where you wind up scraping the front or the back of the trailer. This is especially pronounced if you are building a rather long tiny house. Particularly troublesome would be something like a slightly raised railroad crossing where you're going up rather quickly and then down on the other side. In a shorter tiny house, like a 16' (4.9m) to 18' (5.5m) tiny house, I recommend going with a drop axle for two reasons. First, at those lengths, the house is not going to be that heavy, so drop axles are fine

for handling the weight. Second, you're not going to get the sort of seesaw effect that you would get with a longer house, which is where scraping could occur.

The **weight ratings** on the axles you are using have to match up to the weight of the house. You have to have some idea of how heavy a house you're going to build; the size and the weight of the house will determine how many axles you will need to support your home, both on the road and when it's parked. You can't build a 28' (8.5m)–long tiny house on a single axle, because you're not going to find a trailer that can support the entire weight of the house on a set of two tires. For a tiny house of up to 14' (4.3m), you may find single axle trailers. A 20' (6m) tiny house trailer will likely have two 5,200 pound (2,300kg) axles allowing for a total weight of 10,400 pounds (4,700kg). Twin 7,000 pound (3,100kg) axles will allow for a 14,000 pound (6,300kg) house, and you might find those on a 24' (7.3m) trailer. Beyond this, the trailer will likely include a third axle to handle additional weight. Whatever the size of your trailer, it's not a good idea to max out the weight. Instead, you want to stay below the maximum. You should never exceed the maximum: this can lead to failure of the axle(s) and/or problems with stopping and handling. This is very dangerous and should be avoided at all costs. More axles means more potential problems, too, because there are more tires that can fail, more brakes that will need to be fixed, and more complexity in general. I have seen houses built on repurposed trailers that clearly exceeded the trailer's carrying capacity. This is unsafe at best and reckless at worst; this kind of house can put others at risk when it is being transported on public roads.

Trailers also differ in the **hitch**, the way that they connect to the tow vehicle. There's a standard hitch, which uses a ball that goes into the receiver on the trailer. There is also something known as a gooseneck trailer, which comes down from the top and fixes somewhere closer to the rear axle of the tow vehicle. This helps distribute weight better and allows for a higher towing capacity. If you have a gooseneck trailer, know that towing will only be possible with a limited subset of large trucks, since the hitch connectors for gooseneck hitches are not below the back bumper but rather in the bed portion of the back of the truck.

When you are towing a large and heavy object like a tiny house, you'll eventually need to stop it. **Electric brakes** and a **brake controller** are essential here. Read more about them on page 139.

Trailers and Insulation

Trailers are predominantly constructed out of either steel or, less commonly, aluminum. A trailer is made up of a skeleton framework that is welded together to create a rigid grid construction. That construction has a certain depth and thickness, and you should utilize that depth and thickness to insulate the bottom of your house. This is particularly important in tiny houses on wheels because the space underneath the house gets a lot of airflow. If you are in a hot or cold environment, that airflow can cool down or heat up the bottom of your house. Insulation on the flooring of your house becomes very important, since the floor is built on top of the trailer.

You can further enhance the closing-in of the framework of the trailer by adding flashing. Flashing is just a piece of sheet metal that's applied to the entire bottom of the trailer. I opted for this on my tiny house, and I like it for a variety of reasons. If you think of the structure of your trailer as a skeleton, then the flashing is like the skin on the underside of that skeleton. Since the top of that skeleton remains open, it creates a "tub" in the framework of the trailer, which is what you subsequently fill in with insulation. Flashing also serves as a very good vermin barrier for mice, rats, and other critters. They cannot

When you can't really go wider or taller, you can always go longer. I love the siding on this double-axle house.

A very long 36' (11m) triple-axle tiny house with a fully integrated porch. This one is by Perch & Nest tiny house builders. This is about as long as they come.

chew through metal, and therefore it is an excellent way of protecting what might otherwise be the soft underbelly of your house. It's also true that when traveling down the road, rocks can kick up and water can splash up, and these things can much more easily get into your house if you don't have flashing in place. If you choose to use flashing, the insulation layer sits on top of the flashing and below the interior flooring. You can use a variety of different insulation products to fill in the space (see page 105). If you don't choose to use flashing to hold the insulation in, your only option is to apply spray foam insulation. Many people go the spray foam route, but if you aren't planning on insulating the interior of the house that way, it doesn't make sense to use spray foam exclusively on the underside of the house.

Note the flashing on the bottom. This creates a natural space to fill with insulation prior to closing it up with a layer of plywood on top.

Closing up the now-insulated hollow space on the floor of the trailer.

THE JOY OF TINY HOUSE LIVING

WHERE TO REST YOUR HEAD: LOFT SLEEPING

Where and how you sleep in a tiny house is—you guessed it!—another tradeoff. That's not to say that sleeping needs to be in any way uncomfortable or complicated in a tiny house; it's just something that takes a little bit more consideration than it would in a larger structure. Since you have a limited amount of floor space, most tiny houses have sleeping areas in a loft. This is clearly the preferred option, although it's not always the case. Loft sleeping comes with a couple of inherent disadvantages.

First of all, you're limited to the overall maximum height of 13' 6" (4.1m), so your loft is likely not going to have much **headroom**. Something like mattress thickness can come into play here, because if you use a 12" (30cm)–thick mattress, you are going to reduce the amount of space available above that mattress. I opted for a 6" (15cm) mattress for my tiny house. That might be a little too firm for many people, but it works great for me, and I appreciate the extra headroom. If you want more cushiness, an 8" (20cm) mattress is a good compromise. In other words, go with the thinnest mattress you find comfortable, and remember, hitting your head when you sit up is

less comfortable than a mattress that is firmer than you are used to!

Second of all, you'll have to solve the problem of **accessing** the loft. Access is either provided by way of a ladder or stairs. I use an antique library ladder to access my loft sleeping area. Some people are not comfortable with using a ladder, though, and opt for a staircase. This is somewhat tied to your physical ability and your planned use of the bed area. If you're someone who gets up several times per night, a ladder may not be the right choice for you. Stairs take up significantly more space than ladders, but you can adapt the stairs to create copious storage space.

There are several other factors to consider when it comes to loft sleeping.

This spacious loft is in a rather large tiny house. You won't necessarily have so much room.

ROOF SHAPE: The shape of the roof plays a large part in determining the nature of the loft sleeping area. If you have a typical gable roof, you're creating almost a right angle where the sides of the roof come together above your head. This definitely creates enough space for one person to sleep comfortably in the middle of the mattress. However, there isn't much headroom to either side, so sleeping two people is possible, but might not work for you, depending on your comfort level with small spaces. Dormers (see below) can help because you can bump the space out, but there are also other

A basic yet elegant design, this tiny house has a shed-style roof that features single-level living (no loft).

ways to shape the roof to maximize the headroom in the loft. One option is a shed roof, which slopes from one side of the house all the way down to the other side of the house (slope angle can vary). If you put the head of the bed on the higher side of the slope, or if you sleep perpendicular to the rise of the roof, you can maximize the feeling of space.

Dormers are bump-outs to the more angled shape of the actual roof. Raising the roof for these sections allows for more interior living space.

DORMERS: The low ceiling of most tiny house lofts can be problematic for claustrophobic folks. One way to mitigate this is through the use of dormers. A dormer is an addition to the roof structure that adds a new, protruding roof section to an otherwise sharply angled flat roof surface. This creates a little bit of additional wall space and ceiling height. It also provides the possibility of including windows on the sides of those walls that you're now creating.

SAFETY AND EGRESS: Lofts can become problematic for safety reasons. A fire or other emergency on the main floor could trap you in the loft. Make sure the loft has a separate exit. It doesn't have to be a perfect or easy way to get out, but it has to be there. In most cases, this will take the form of a window that opens up wide enough to allow exit. According to the International Residential Code (IRC), an egress window needs to be 5.7 square feet (0.53 square meters) on non-ground floors, or 5.0 square feet (0.46 square meters) for the ground floor. Make sure your window fits the bill.

A skylight in a roof like this offers another means of getting out of the house in the case of an emergency.

Oakley paying a visit to one of my tiny house builds. Make sure your tiny house is pet friendly for both your sake and your pet's.

CHILDREN AND PETS: If you have a loft and ever anticipate children entering your house, put railings on the edge to prevent accidents. If kids will sleep in the loft, railings are absolutely essential. When it comes to pets, most of them won't be able to climb a loft ladder, so if you want them to spend time in the loft with you, you'll need stairs.

WINDOWS AND LIGHTING:

Good lighting is essential in a loft. A lack of light in a loft makes it feel like a coffin, but if you have windows lining the edge and the back area of the loft, it can feel spacious, bright, and airy. Windows also help with ventilation. The loft area under the roof tends to be one of the warmest areas in the house, since heat rises. In the summer, you need to let some of that heat out and open the windows to create cross-ventilation. Even in the winter, the heat can get stuffy. This is especially true if your heat source is one that doesn't allow you to control heat very precisely (such as a wood-burning stove).

Windows up high let in a ton of light. It makes all the difference in creating a space that feels open.

John and Fin Kernohan's beautifully and well-lit "Firehouse Station No.9" tiny house.

A downstairs bedroom with a queen-sized bed. A mattress would make this more comfortable.

The Loft Alternative

Even considering all the tips and tricks to maximize space in a loft, there are some people who are just really put off by the notion of sleeping in a loft, and you might be one of them. It is possible to put a bedroom downstairs in a tiny house, but it is going to take up major floor space. Make sure that you get a long enough trailer to accommodate the arrangement. If you go for a full-size or a queen-size mattress that's aligned lengthwise with the trailer, you'll have an area where you can walk past the mattress on one side, but the other side of the mattress is going to have to be pushed up against the wall. If two people are using the bed, one of them is going to be on the wall side, and it's going to be more difficult for them to enter and exit the bed. With the proper furnishings, you may be able to wrangle a fold-down bed to save space.

You'll have to clamber into this ground-floor bed. This is my daughter's bedroom in the first tiny house I built.

LET THERE BE LIGHT: WINDOWS

Windows and the use of light in a tiny house are something that in many ways can make or break the house. When you have such a small space, you want to let in as much light as you possibly can. For this reason, you want to make sure that you're not skimping on windows. I said earlier that the trailer, which is ultimately the foundation of the mobile tiny house, is an item you shouldn't skimp on. Windows are another significant part of the house where you should use as few shortcuts as possible. Let's look at various considerations for windows.

A window fully sealed in place and installed on an unsided house.

FRAMES: Window frames are made out of different materials, and those different materials have different properties and different price points. Four common options include fiberglass, PVC, wood, and metal (typically aluminum). Which of these are best for tiny houses? Strike metal off your list right away, because metal is a very effective thermal bridge, meaning that if it's cold outside, that cold transfers very quickly, efficiently, and readily to the inside of the window. This is not only bad in terms of temperature, but it also tends to create condensation. Wood is a perfectly good material for frames; typically, wood frames are somewhat more expensive. PVC frames tend to be the least expensive, and fiberglass is a sort of hybrid between a wood and PVC frame. How you source your windows will likely be the biggest factor in determining the final cost. I have gotten some amazing windows through Craigslist.com for pennies on the dollar. If you can work your find into what you are building, that is a great way to go. If that's not an option for you, then buy from a single manufacturer. Simple picture windows of an average size will be around $100. Different opening and quality options will raise that price to potentially around $200. I installed ten rather high-end windows into my original tiny house, and they cost me around $3,000 total. They've been great, but I likely didn't need to go quite that fancy.

This good-looking window is nicely integrated into the side of the house.

Consider spending the money on double-pane or even triple-pane windows; you'll reap the benefits in energy efficiency.

PANES: Historically, most windows were single pane, which means a single piece of glass between you and the outside world. Over time we've created much more effective ways of creating glass panes, and residential-style windows are now almost exclusively made as double-pane windows. That means that there are two pieces of glass separated by a space, which acts as a layer of insulation. The space is filled with argon or krypton gas (ineffective against Superman), which further enhances the properties of the window insulation and blocking harmful rays. Because modern windows tend to be very energy efficient, adding a large number of them does not significantly detract from the energy efficiency in a tiny house. For those who want the ultimate in energy efficiency, though, there are triple-pane windows made of three pieces of glass with two spaces. A triple-pane window does a lot to diminish noise, too. All this efficiency and insulation comes at a cost, though: these are your most expensive window options.

QUANTITY AND PLACEMENT:

You're going to want several windows in your house. The thing to keep in mind with a tiny house is that proportion plays an outsize role in everything that you do. I've seen people pop a rather large window into a tiny structure. This can certainly be done for effect, and it can look pretty good, but if you're going to arrange a number of different windows around the exterior walls of your tiny house, you will want them to be roughly in proportion with the house itself. That means that typically you will go for slightly smaller windows than you would in a standard structure.

Usually windows are best kept in proportion to the house size: a small house means small windows.

THE JOY OF TINY HOUSE LIVING

PICTURE WINDOWS:

Windows that do not open, which just let in light and provide a view, are called picture windows. They can be rather large, often consisting of one big piece of glass. I have used picture windows in houses where there's a particularly picturesque view to take advantage of.

This gorgeous picture window is a great thing to wake up to.

FUNCTIONAL WINDOWS: Windows that open come in many forms. Casement windows are hinged on the side and fold out sideways; awning windows fold from the bottom out. Awning windows are good for use in inclement weather conditions; you can keep the window open, because rain will generally just hit the glass and drip down off the edge. This function is also good for maintaining ventilation during bad weather. There are also windows that are separated into top and bottom sashes, where one or both of the sashes can be opened. All other things being equal, casement and awning windows tend to be more expensive since the hardware and mechanics to operate them are a bit more complex and involve hinges and hand cranks.

SCREENS: If you're in a situation or climate where there is going to be an abundance of insects that could get into your house, you're going to want screens on any windows you plan to open.

EMERGENCY EXITS: Windows are also an essential means of getting out of the house in an emergency (as discussed in the lofts section, page 64). There should certainly be a window in the loft that you can use to exit the house in case of an emergency, but there should also be an alternative exit on the ground floor, besides the front door. So don't make all your windows picture windows, and make sure at least one of your windows is big enough to climb through.

THE GREAT OUTDOORS: EXTENDING YOUR SPACE

My girlfriend lives in an apartment in New York City, and her place differs from my main residence in that she does not have a balcony. I've come to realize how important it is to me to have that extra living space that extends outside. Extension of space especially applies to tiny houses, since the interior spaces are smaller. When the interior space is small, it makes sense to enhance and utilize your exterior space as much and as efficiently as possible. Think of the outside of your tiny house as an extension of your living room. Hopefully you can put your house in a place where it's pleasant and agreeable to be outside most of the year, whether it's in someone's backyard, a campground, a trailer park, or a similar location.

I have a **fire pit** next to my tiny house that serves as both a seating area and also a means to cook. A fire pit can be as simple as a few large rocks assembled in a circle, which costs nothing, or as complicated as the manufactured steel kind that can run a few hundred dollars. Get what you like and what you can afford. I don't cook all my meals at my fire pit, but it's nice to have the option. Regardless of where my meals are cooked, I utilize the seating area

An excellent, useful porch. This is space added, not space wasted.

The outside surroundings of my tiny house.

to eat most of my meals as well (weather permitting). If you're fortunate enough to find a place for your tiny house where there are nice views to be had, then an outdoor seating space makes even more sense.

A **porch** on a tiny house is another feature to consider; it's a very small bit of extra outdoor space, but it can make all the difference. My tiny house has a porch, and I wouldn't trade the two feet of porch space for interior space at any cost. I love sitting on the porch and do so quite often. When it's raining outside, the porch is great because you can enjoy the outdoors without getting wet. Another great aspect of a porch is that it gives you a place where you can

shake the snow off your boots (lucky you if snow is not a thing where you live). You can remove outer layers of garments and even take your shoes off before entering the house. Once you get into the house, space is limited, so utilizing the exterior as a makeshift foyer can be helpful.

If you opt for a porch, know that you will have to dedicate space for it that comes out of the overall length of the trailer: a porch is typically not added on to the end of a trailer, but rather is part of the trailer. Therefore, porch space means less interior space. If your tiny house is mobile, make sure to remove your rocking chair and plants before you head out on the highway. If your house is likely to never travel, you may be better off designing a deck structure on the exterior of the house. This structure will look like it is part of the house even though it is unlikely to be physically connected. This gives you the flexibility to move the house and have a deck/porch at the same time without giving up the ability to maximize interior space.

Some people also opt to put their **showers** outside. Usually this is for small structures like campers or traveling tiny houses. I get a lot of inquiries in my business about outdoor showers; they seem to be popular. They are a way to save interior space on something that you don't actually spend very much time in.

As I think you can tell, what I'm getting at here is that location and environment play a key role in so many aspects of your tiny home. I understand that given the legalities and the problems with finding space for tiny houses, it's not always going to be possible to place your house in an area where the exterior space is suitable for frequent use. But if you are fortunate enough to be able to take advantage of your surroundings, then by all means, make that outdoor space work for you.

Installing an outdoor shower is one way to save space inside and also enjoy the outdoors.

This is right outside my front door. There's no need to hang out indoors all the time if you have this.

THE JOY OF TINY HOUSE LIVING

ANDREA BURNS

Andrea: I'm Andrea Burns, the owner/operator, builder, and CEO of Andrea J Burns Inc.; I live in my tiny house, Tomato Box. I wanted something that had a lot of natural light and a good bit of airflow. I wanted to make sure that if I were going to an environment that small that it was a healthy environment.

Chris: *Now that you're living in a tiny house full-time, what are one or two things you've discovered?*

Andrea: I thought I would miss my TV after I put it in storage, but I have a laptop and a phone, and I can watch Netflix on that. I have all the entertainment I need. I don't spend a lot of downtime trying to be simply entertained. I watch TV once or twice a month. The rest of the time I'm either working in a job, working on the house, or making art or something else. I don't spend a lot of time passively consuming things.

Andrea Burns on the porch of her tiny house.

What a nice entryway. I'm always a fan of the unusual.

Chris: *What's your ultimate goal in the tiny house world?*

Andrea: I started doing it for myself. I felt like it was a lifestyle I would enjoy, but then I spent five years working with underserved groups of people. I started out working with the elderly in nursing homes, and then with kids and after-school programs, and then with those who are just trying to avoid being homeless. That's a hard life, and I realized that we were sitting on a gold mine—not in the money-making sort of sense, but in the "This is awesome!" sense. Tiny houses could alleviate a lot of the stress involved. I was homeless for a minute; I was couch surfing with friends, and I underestimated the amount of stress involved.

Chris: *What is the most common question people ask you about living tiny?*

Andrea: Since it's contrary to what we're taught growing up, I get a lot of questions that revolve around whether it is something a person can realistically do. And I get a lot more questions from women than men, because women are taught from day one that they can't do things, so they come in, they see me, and they go, "Wait, you live by yourself?" That's the first thing they're told they're not allowed to do. I say yes, and they ask, "You built this yourself?" and I say yes again. They're trying to wrap their head around the question, "Is this realistic for me as a woman or am I just dreaming?" And the answer is that it is realistic.

THE JOY OF TINY HOUSE LIVING

A great door is a good way to make a tiny house pop.

Andrea: The next step is the nuts and bolts of it. I've decided I can do this. I've gotten rid of the preconceived notions. I'm going to do this. How do I do it? Nobody taught me how to do this. It's a lot of step-by-step stuff. I try to emphasize that there's no one way to do it. What we're taught in this culture is that there's one right answer for everything. That if you want to get a good job, here are the steps to getting a good job, and it always involves college. It always involves grades. Except in real life, it doesn't. These ideas in our heads, these notions that we're carrying around, they're baggage, and if you're going to get in a tiny house, you have to let go of your baggage. People are looking for the step-by-step instruction manual for going tiny, and there isn't one.

Chris: *That's why I'm writing this book.*

Andrea: I'm glad you're writing one. It's like a whole matrix of checkboxes; well, you could do it this way, or this way, or this way, and they all work, but they work for different circumstances. One of the things I'm very passionate about is critical thinking. It's considering your circumstance to see what fits you best. The trouble is, you get somebody who's in a different circumstance, and they try to match what you did step-by-step. We have to get a little bit better at doing our own research and doing our own thinking.

CONSTRUCTION

Having a plan is great. Now it's time to execute! There are some key decisions you need to make before the first board gets cut. If you are outsourcing your build, this section is more for reference, but you should still take an active role in the construction. For those forging ahead and building on your own, there is beneficial information here on all the main construction aspects you'll encounter along the way to building your tiny house.

HOW WILL YOU PAY FOR IT? FINANCING

All this talk of tiny houses so far hasn't included any in-depth discussion of how you are going to pay for your tiny house. Maybe you're in the fortunate position where you have ample amounts of cash laying around that you can use to pay for the house outright. Excellent, good for you! However, most people have to get creative about funding their tiny house. Tiny houses aren't necessarily suited for mortgages, but that's not terribly important, since you may be interested in tiny houses in the first place because you're not interested in a long-term payment structure that follows you around for 30 years. There is middle ground between the cash purchase option and the 30-year mortgage option, though; read on.

Make it a policy to always save your change for your tiny house project. Over time, the money will start to add up.

It is possible for most people to get some form of **personal loan**. Banks grant loans for all sorts of reasons, whether it's to cover medical costs, to put an addition on a house, to start a business, or even to buy a boat.

Larger builder operations often now provide **financing** as well, usually in the form of a payment plan. It's in their interest to do so, and it could potentially work for you. These financing plans aren't offered by the builder per se, but rather by third parties with whom the builder works. That way the builder always gets paid for their work and their product, and

These repurposed iron pipes make a very cool bookshelf.

the banking entity handles the financing part with the purchaser.

Another thing you can do is build your tiny house in **stages**. You can spread the building process out over a few years, starting by buying the trailer, then acquiring pieces to build the house over a period of time. Doing it this way allows you to spread the cost out over an extended period and makes it easier for you to incorporate the house into your overall budget.

By building in stages over a long period of time, you can take advantage of sales and other opportunities to spend less on materials.

The other advantage of building gradually is that you'll be able to source materials more effectively. You can shop around and get the best prices for the materials that you need, sometimes finding great deals on Craigslist, on eBay, at yard sales, and so forth. Building in stages goes well with the use of repurposed materials, too—the longer you take to build, the more time you have to get the materials, and the more likely you are to get those materials at a reasonable or reduced cost.

Friends and family are another means that people can utilize to finance a tiny house. Do you have someone who is willing to lend you money? You may find it easier to borrow money from them than to take out a more traditional personal loan through a bank or a credit provider. Do you have someone who can help you with a particular aspect of your tiny house, maybe an electrician friend to help you with wiring or a woodworker to help you raise the walls? Get creative and use the connections and resources available to you. And if someone supports you with time or money, make sure they get an opportunity to spend time in your house. Some people are also exploring resources like crowdfunding; your situation and abilities determine whether or not this applies to you.

Last but not least, there's always the option to **save up** for your tiny house. Put all your change in a piggy bank, skip the vacation trips, cut out that weekly Starbucks habit, stop smoking—the money will accumulate. Keep in mind that you don't necessarily need to save up the entire cost of the house before beginning construction, unless you're set on purchasing the house straight out. You may be able to start building the house once your tiny house fund has sufficiently grown.

If you cut out your expensive coffee habit by sticking with homemade brew, you may be able to save the money you need to fund your tiny house. Then you can enjoy home-brewed java from the comfort of your tiny house yard, like I am doing here.

All the different aspects of construction aren't for everyone. Will you be comfortable climbing up a ladder and drilling screws into your roof? Know yourself and work within your limits.

CAN (AND SHOULD) YOU CONSTRUCT YOUR HOUSE YOURSELF?

Personal ability to construct a house is a very subjective topic, and it involves your assessment of your own ability to build a house. Realize that a lot of other people have built tiny houses with little to no background in house building—it *is* possible, so don't rule it out straight away. Declaring, "Well, I've never built a house before, therefore I can't build a house," is not the right attitude to take. That doesn't mean everyone *should* build their own tiny house, though.

Part of determining this has to do with your affinity for risk and how risk plays a role in the rest of your life. Are you someone who avoids risk at all costs, or are you someone who is open to it? Are you someone who dives headfirst into things you've never done? Examining your answers to these questions will help you figure out whether you are mentally able to take on a project like this. There are a few other specific elements to consider as well.

Even if you're physically capable of all the tasks necessary to build a tiny house, there will be times you will greatly benefit from a helping hand.

PHYSICAL ABILITY: When constructing a tiny house, some aspects require a certain amount of strength, or may need to be done by two people. Being able to lift at least 50 pounds (22kg) by yourself is definitely helpful. An example of something that really requires two people is putting in a large picture window, which is both heavy and awkward to manage on your own. Work involving the roof area is also good to have help with, since having someone hand you items/tools is way more efficient than constantly going up and down a ladder to get them yourself. Building a tiny house all on your own is possible—I have done it. But at the minimum, I advise you to have someone on hand as a backup in the physical assistance department. Even if you could do a thing yourself, you will save a lot of time and effort just by having someone lend a helping hand once in a while.

KNOWLEDGE BASE: If you've never built a house before, it's not possible to know everything you need to know about everything you'll need to do. Whether you need answers related to plumbing, electric, or the right way to set a screw, there are plenty of subject matter experts available online. I answer people's questions all the time via my website (*www.tinyindustrial.com*) or my Facebook page (*www.facebook.com/tinyindustrial*). YouTube and Google searches are other ways for you to acquire a vast amount of essential knowledge that wouldn't have been possible 20 years ago. The information is out there; you just have to find it.

Another thing I want to stress about knowledge base is that building your house is not a race. Setting your own pace will make the process comfortable instead of daunting. When I built my first tiny house, there were aspects of the house that I didn't know how to tackle. I went through a rather lengthy process at times to figure out how to do certain tasks, such as installing plumbing. When I needed to do research like this, I simply stopped work on the tiny house. I had that luxury; I wasn't paying for space where I was building the tiny house, so it could just sit there and wait for me to reach the confidence level in a given subject that I needed in order to progress to the next stage.

RESOURCES: What other resources do you have? Do you have a family member who's in construction? Is there an electrician in your circle of friends? Think about all the people that you could involve in your tiny house project. People are generally glad to help, and they're often excited about tiny houses. It's a topic that resonates positively with many people, so chances are you'll be able to recruit helpers. On the flipside, if you're rather isolated where you live or new to the area and don't have people you can call on for help, you might be biting off more than you can chew by trying to tackle all the construction yourself. Be realistic.

PARTIAL SOURCING: One other path to explore when deciding whether or not to construct your house yourself is to consider getting a shell made. Tiny house builders make full end-to-end tiny houses, but some of them can also build a tiny house shell, which is essentially a trailer with framing, walls, and a roof (either temporary or finished). You can often choose a la carte features as well, such as whether you want the electric roughed in or the plumbing done or the windows added. A lot of tiny house builders (including me) are open to working with you on what you want out of the shell.

Keep in mind that the more things the builder does, the more expensive it's going to be, but it's still less expensive than having the entire house built, and if that's what it takes to make you comfortable with construction, then that's fine. A shell is a hybrid option that is a good middle ground where you feel ownership over your home but also confident in its structural integrity.

To give you an idea of pricing of shell versus completed house, a shell is typically going to be a bit less than half of the cost of a completed house. Using the image on this page as a guide, the most basic shell would be around

If you were to purchase a "shell" from me, it might look something like this. All the hard work and heavy lifting has been done for you. You get to close it up and finish out the build. It saves a lot of time.

$14,000 (including trailer, framing, and sheathing). I sell a sealed shell of this house (trailer, walls, windows, door, and roof) for around $20,000 (dependent on specific features). This house with a finished interior, siding, insulation, electrical, and plumbing runs between $38,000 and $44,000. Use these number as a relational guide only: larger houses will cost more, but the relationship between the numbers will progress in a similar fashion.

One last note about building yourself. Understand that your level of knowledge will directly impact the safety of the structure. Safety is incredibly important because not only is this your house that you're going to be spending a lot of time in, but it's also going to be towed down the public highway. Other people are involved, people who share the road with you, people who could be injured by any mistakes that you make in the construction of the house. You may be willing to take risks for yourself, but do not take them for others.

HIRING A CONTRACTOR

If you decide to outsource your tiny house's construction either fully or partially, whom you outsource to is an important consideration. The rising popularity of tiny houses has led to many general contractors, people who do some construction work, and people who build regular-size houses or other structures jumping on the tiny house bandwagon and saying, "Hey, I could build a small house like that." In theory, this is fine, and there are many good examples of people doing this successfully. But the thing that many general contractors miss is that building a tiny house is very different from building a larger house. I don't mean in terms of raw skills or in simple sizing; I mean in terms of creating a livable space and a sound structure that will serve its purpose.

Since most tiny houses are built on wheels, they must be able to travel down the highway at high speeds. Dragging a house along at 60 miles per hour (100km per hour) is going to stress it in a way that it would never be stressed if it sat quietly in a lot. This is what makes tiny houses so different from standard houses, and many contractors don't necessarily adhere to the more stringent standards that a tiny house will require.

Expect to pay around 20 to 25 percent more to hire a company that specializes in building tiny houses versus someone who is a general contractor building a tiny house on the side. The extra money is for the company's knowledge and understanding of the fundamental differences between building a house on a foundation and building one on a trailer. General contractors also are unlikely to have the insight into the various plumbing, roofing, and siding options,

Hiring someone to build your house? There is a lot to consider.

etc., that make the most sense for a tiny house. I can usually tell when I'm in a tiny house that was built by a general contractor. These tiny houses will often have a flush toilet, a standard size fiberglass shower, and vinyl siding. There's nothing inherently wrong with that—it's just that there may have been better options to implement.

General contractors may have a lot of home-building experience, but you'll need to be sure they can adapt their experience to the needs of your tiny house. For example, bricklaying experience isn't helpful for a tiny house that is meant to be mobile.

So how do you protect yourself against an "amateur" approach by a professional **general contractor who does not specialize in tiny houses**? There are many things you can do:

- Find out what tiny houses the contractor has built, and how many.

- Take a look at their previous work, if there is any way to do so.

- Ask for references.

- Try to choose someone local; that way, you can keep track of what they're doing and how they're doing it.

- Gauge how willing they are to accommodate your needs around the build. Are they open to building what you want, or are they trying to sell you a house that they already have, or a model that they're comfortable building?

- Figure out how they are sourcing the trailer. As we discussed earlier (page 57), a good trailer is essential, and the contractor's answer to this question can really reveal what you are going to get for your money. If they're going to cut corners on the trailer, then they may cut corners on all sorts of other things (and perhaps not tell you).

- Ask a ton of questions about the process and the materials. I can't stress this enough: educate yourself on the types of materials that work best in tiny houses, and make sure your contractor is using them.

If you're going with a reputable **tiny house manufacturer**, you can rest easy about a lot of the things that could be concerns with general contractors, and reap some other benefits as well:

- Usually, if a company has grown to the level of being a manufacturer, they've got a track record in the business, and it should be pretty easy to ascertain the type and quality of the work that they do. Know that their work is likely to be more expensive because of their specialized expertise and reputation.

- You may be able to get an RVIA certification for your tiny house, which is the seal of approval from the Recreational Vehicle Industry Association indicating that your house has been built to a particular set of standards.

- A manufacturer may be better able to help you obtain financing for your tiny house.

- Insuring your house may be easier, because insurance companies can look up the model of a tiny house and get the exact costs/value of that particular house. The manufacturer's other customers may also be able to recommend insurance agents.

The bottom line is that when you outsource the work of building your tiny house, make sure you do it carefully and put a lot of thought and effort into the process. Unfortunately, I've seen tiny houses that were constructed so poorly that the house couldn't be lived in, and certainly was not safe to haul down the highway. Don't let this happen to you; do your research and make smart decisions.

If your contractor is just going to grab any old trailer off the lot, they may not be a good choice. Don't be afraid to speak up for what your home needs.

BARE HANDS AREN'T ENOUGH: TOOLS

If you are going to do some or all of the construction yourself, you're going to need some tools. Tiny houses can be built with a pretty minimal toolkit. Don't think that you have to go out and spend $1,000 on all sorts of things. Keep in mind that you may be able to borrow the vast majority of the tools you need as well. Unless you plan to build more things, you may never need some of these tools again after your house is done. The pictures in this chapter are of my own well-used tools. They are nothing fancy, but they work just fine, and I have built multiple houses with them. You may find yourself gravitating toward other tools as you work and gain more experience, but these are a great starting point.

Beyond the specific tools we'll go over in this section, you will also need some other **miscellaneous things** like clamps, saw horses, utility knives, caulking, and so on. Since this book is not intended to be a hands-on how-to book, I won't delve too deeply into these peripheral items. Suffice it to say

that you will use these on an as-needed basis, and they are not going to break the bank.

Tools broadly fall into one of two categories: **hand tools** and **power tools**. Power tools in general are a bit more expensive than hand tools, but unless you have some personal reason to only use hand tools, power tools will make your life a whole lot easier, and I recommend spending the money on them.

Power tools come in two subcategories: **connected/wired** tools and **battery-powered** tools. When I grew up, there was no battery option, so I tend to gravitate to connected tools. I see more and more people using battery-powered tools, though, and I have started to supplement my toolkit with some battery-powered tools, too. Battery-powered tools are excellent because you don't have to worry about cords getting in the way and where to plug them in. They deliver a good amount of oomph these days, too. The downsides are that they cost more to purchase, you can't work when the batteries are depleted, and the batteries don't last forever.

The Bare Minimum

So, what are the tools that you will absolutely need? Here are the tools required for tiny house construction:

CIRCULAR SAW: because you'll need to cut things

POWER DRILL WITH BITS AND ATTACHMENTS: because you'll need to attach things together

MISCELLANEOUS: straight edges, pencils, a level, a hammer, screws, nails, etc.

Circular saw. This is useful for cutting things in a reasonably straight, unprecise sort of way.

Power drill. My old one finally died, and this one is a relatively new addition to my batch of tools.

That's it! With these two power tools and a few supplementary items, you can build a tiny house. I don't recommend that you proceed with just these two tools, but my point is that there is no need for tons of expensive equipment. You can buy these two power tools for under $100.

Well Worth It

Let me add some other tools that make the job much more manageable, quicker, and more accurate. I consider the tools in this section to be recommended additions to your tool arsenal. These tools combined will cost you between $500 and $1,000, depending what quality grade (i.e., name brand) you go with. With the exception of the jigsaw ($30), the other items all run between $100 and $150. Remember, you can always buy some of these used to save money.

MITER SAW: These are great to make accurate right-angle cuts. They also can cut at angles other than 90 degrees (typically up to 45 degrees). Since tiny houses are often constructed with 2x4s (50x100mm) for the framing, this tool is invaluable for making numerous and exacting cuts to specific lengths. Can you frame a house without a miter saw? Sure you can, but I wouldn't want to.

CORDLESS IMPACT DRIVER: Have you ever tried to screw in a 3" (7.5cm) screw by hand? In case you haven't, I can tell you: it's not fun. Now imagine having to screw in a couple hundred screws like this. This is where the cordless impact driver comes in. It is an electric motorized screwdriver on steroids. It's a recent addition to my toolkit, but I wouldn't want to do without one of these anymore; I screw all my framing together with this tool. Screws create stronger bonds than nails, and are therefore preferable. You can use a good power drill to drive screws as well, which is what I used to do, but a cordless impact driver is way more convenient, since it comes in a small, lightweight format and is excellent for powering in any number of different fasteners. The impact part of this tool also differs from a regular power drill in that it utilizes a hammering mechanism to spin the fasteners into place.

JIGSAW: You will need to make rounded cuts or lop off corners of things. The circular saw is not going to do that for you; it's too big and cumbersome. Enter the jigsaw. This works great on lots of different materials, and you can swap the blades for whatever your current cutting task requires. An inexpensive corded jigsaw will do.

Miter saw. This is an inexpensive one, but it gets the job done.

Cordless (battery-powered) impact driver. It's small and versatile, yet also very powerful at driving screws.

Jigsaw. One of these gives you some more precision in cutting and is very good at cutting curves and making rounded cuts.

AIR COMPRESSOR AND NAIL GUNS:

I use this combo for interior work, mostly. The compressor fills an air chamber with pressurized air, which then can be used to power various devices. The main thing I use a compressor for is to use a finishing nailer to shoot very fine and thin (headless) nails. These are great for fastening pieces of trim or interior paneling. On my tiny house, I also used a framing nailer to secure the cedar siding to my house. This is a much bigger gun than the finishing nailer, and it shoots rather large nails efficiently, quickly, and repeatedly. You can even get a car tire filler attachment for your compressor, too; it mimics the machine at the gas station that puts air in your tires. This can come in handy and will save you the occasional dollar!

Air compressor. Compresses air, which is then used to power various things you can hook up to it.

The gauges on the air compressor show you how much pressure is in the tank and how much you are delivering to whatever you have hooked up to it. You can dial the pressure up or down based on the needs of the device.

Finishing nailer. Use this for attaching interior trim in such a way that you can barely see. Fast and efficient, it beats gluing stuff on the walls.

Framing nailer. This is the heavy-duty version of the finishing nailer. It shoots huge nails into things you want to attach together.

You can see the nail in there. This is one of the more powerful tools you may want in your arsenal.

Handy Extras

There are a few more tools that I love having on hand. You can expect to spend about $300 total on these three tools.

OSCILLATING MULTI-TOOL: This thing is excellent for achieving those nearly impossible cuts. It works by essentially vibrating a blade back and forth. That blade is typically a relatively narrow one. When you have to trim something off a piece that has already been installed, this is a lifesaver. It provides small, precise cuts that no other power tool can make in that situation.

Oscillating multi-tool. You attach one of various kinds of blades to the front of this and it vibrate-cuts its way through things. It's great for getting you out of a tough spot during a build.

The multi-tool getting ready to do what it does best in its natural construction environment.

RANDOM ORBITAL ELECTRIC SANDER: Sanding with sandpaper is okay once in a while. You can use a file for some other jobs. But for larger-scale smoothing, an orbital sander is ideal. You can stick various grits of sandpaper onto it, and the vibrating action takes most of the elbow grease out of the equation. I reach for this tool frequently during construction.

Electric sander. This will save you a lot of elbow grease.

The sander is great for smoothing out a broad wood surface like this. Progressively finer grits of sandpaper on the sander lead to a buttery smooth finish.

TABLE SAW: You can make long cuts with a circular saw and with a jigsaw, but getting those cuts perfectly straight is not possible. A table saw allows you to make perfect long cuts, since it's designed to keep the blade straight as you run a piece of wood through it. The fence on the top of the table saw gets locked into place at a specific distance from the blade. That distance determines the width of the cut. Whatever you push across the top of the table saw is cut precisely to that width. Safety should always be a top concern, but this is especially true with a table saw—it's one of the more dangerous tools I've included in this section. This is mostly because you are pushing an item into a cutting blade. Don't slip. Wear eye protection (always). Stay clear of the blade. For smaller, narrower cuts, there is a device called a push stick that you can use to push the object through the blade. It's an extension of your hand, with the difference being that if the push stick gets in the blade, it's no big deal. If you choose not to use a push stick, get ready to call for an ambulance with your undamaged hand.

Table saw. Make sure you have one of these when you absolutely need to cut in a straight line.

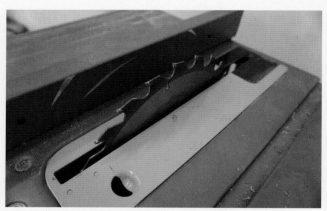

You can raise and lower the blade to cut to different thicknesses.

The table saw fence is where you set the width of your cut. This is set for 2" (5cm). You can see it in the small window.

This is the push stick. Alternatively, you can use a scrap of wood to push things through the blade. Either way, don't let your hands get near the blade. The blade will always win.

THE WALLS GO UP: CONSTRUCTION

No matter what level of physical involvement you have decided you are going to have in the construction of your house, I still encourage you to be very much involved in the construction process. It's best to know how a house is built, what materials were used, and why certain decisions were made. Later on, when it comes time to fix something, modify something, or improve something, having that knowledge is immensely useful for making smart, informed decisions.

There's a typical order to the basic construction of your tiny house, which generally goes like this:

1. Flooring

2. Framing

3. Sheathing

4. Roofing

5. Interior framing

6. Electrical/plumbing

7. Insulation

8. Finishing

This order is not set in stone, but it usually makes the most sense. Let's look at each stage in a little more detail.

So, typically you start with the **flooring**—more specifically, the subfloor. This is what gets built up onto the trailer platform, and this is where the walls will go up as a part of the framing.

Next, there are three ways that you can **frame** a house:

2x4 (50x100MM) FRAMING: Also known as stick frame construction, this is by far the most common framing method. Virtually all private residences and some multiunit dwellings in the United States are constructed via this method. The process is to frame out the skeleton of the structure using this ubiquitously available lumber dimension. This is the most widely accepted norm for construction.

This is what basic stick frame construction looks like.

METAL FRAMING: Metal framing involves steel beams utilized in place of wooden 2x4s (50x100mm). These are not solid steel beams, but they have a specific structural rigidity. They are, at times, somewhat lighter than wood beams, and are cut and usually connected rather easily with appropriate screws.

Here you can see the metal framing in place of where traditionally there would be wooden 2x4s.

A tiny house fully framed in metal. The next step: putting up the sheathing on the exterior.

SIPS: SIP stands for structurally integrated panel, which is a panel that consists of two sheets of plywood sandwiching a foam core. SIPs provide structural rigidity as well as insulation and sheathing in one step. This is an immense time-saver over building a house using a traditional stick frame construction, because it cuts out the extra work you would typically have to do with 2x4s (50x100mm), including sheathing and insulation. Furthermore, companies that will cut these SIPs for you can do so according to an architectural plan. The precut panels then fit together very quickly with a minimum of tools. That means you could effectively create a house that has four walls and a roof with five SIPs. Note that there are some drawbacks to SIPs: it can be difficult to run wiring and plumbing through these pre-integrated walls. Understanding and knowing what you're trading off is the key to making an informed decision.

A small-scale sample of a structurally integrated panel (SIP). You can see the white foam core between the pieces of plywood.

The bigger consideration when choosing between these three methods won't be cost; it is really more about labor and flexibility. Wood is the cheapest and most readily available material to use and also offers the most flexibility, since you can build whatever you want and can even change the design on the fly. When you work with wood, though, you have to make lots of cuts and piece the framing together—it's a lot of work.

The other two methods vastly reduce the amount of labor. They are more expensive, but time is money, as they say. You are also giving up flexibility. Many of the metal framing setups are shipped precut and designed to be put together in a certain way to build a specific structure. Windows will be incorporated in certain places and be of a certain size. Rarely would you build from scratch using metal framing. Think of it more like a Lego® set to build a certain kind of house.

SIPs takes this one step further in that your house may only consist of five pieces (four walls and a roof). These will have to be produced and precut by a giant machine and then shipped or picked up on a truck. This means another reduction in labor, but with an even greater cost to flexibility. Everything from windows to where the door goes is all already precut. The advantage here is that you also have all your walls insulated already, although running electrical and plumbing is made much more difficult in walls that are already closed.

To sum it up:

STICK FRAME (WOOD):

- Pros: cheap, readily available, flexible design

- Cons: labor intensive, potential longevity issues (rot, vermin, etc.)

METAL:

- Pros: sturdy, somewhat lighter than wood, quick to put together from pre-cut pieces, termites don't eat metal

- Cons: expensive, inflexible design, thermal bridging (metal is a good conductor of hot and cold)

SIPS:

- Pros: quickest way to assemble a house, light, sturdy, walls are pre-insulated

- Cons: no flexibility once designed, few manufacturers, shipping costs, more difficult to retrofit plumbing and electrical

Before I start framing the interior, I like to put the finished flooring in. You have a nice, big, open space, and it negates the need for doing a lot of cutting to put flooring around the interior walls.

Once the framing is completed, you move on to **sheathing**. This lends rigidity to the house, because once large plywood panels are covering the 2x4s (50x100mm), there is less racking and movement possible. Racking is the notion that something you build won't inherently be strong in all directions. Flexing and side-to-side sway is what sheathing puts an end to. 2x4s (50x100mm) have small and limited attachment surfaces. Those surfaces are not substantial enough to prevent flexing and swaying. When you attach a piece of plywood sheathing over framing like this, you create many more attachment points and, by extension, strength and rigidity. During framing and sheathing, the location of all the windows is determined. This is also when **roofing** occurs (more detail on page 108) .

After the windows go in, you'll want to do the **interior framing**. Your house is not just going to be a big empty box on the inside (probably). You're going to have a bathroom that's going to need closing in, and there will likely be a loft sleeping area that has to be constructed as well.

After the interior framing is completed, you can move on to **electrical and plumbing**. You'll have to decide where sinks, outlets, kitchen implements, the toilet, and potentially the shower are going to go, and then run electrical and plumbing accordingly. Thoroughly test all of the aspects of the plumbing and the electrical system to be sure that it's all done right before moving on to the next stage. Mistakes you make inside a wall are very costly once you've closed it up. Some things are a little bit hard to test; for example, it's hard to test the shower when the walls are open without water getting everywhere, but it does make sense to at least pressure test the lines. I've sometimes capped specific lines to test other parts of the house and then removed the caps later on. It was additional work, but it was worth it to not have to fix anything later on.

Once the utilities have been installed, you'll move into the **insulation** phase. Insulation is covered in more detail on page 104.

After all of this, it's time to **finish**: close up walls, install fixtures, and decorate. Voila, you've got a house!

This is interior framing, since you don't just have walls on the outside of your house.

Aside from my dad and a few other things, you are looking here at 2x4s and plywood. That's mostly what a house is.

POWERING IT UP: ELECTRICITY

There are two kinds of tiny house electrical systems: on-grid and off-grid. There are advantages and disadvantages to both; you'll have to decide which one suits your needs. We will examine both in detail in this section.

On-Grid

An on-grid system, in which you draw power from the local utility company or other established source, is modeled after a traditional house power setup. The only difference between a tiny house and a regular house is that the electrical hookup is probably a less permanent one, since the tiny house is usually made to move. Therefore, you have to run a wire to the house, instead of the connection being permanently attached.

There are standardized **hookups** that recreational vehicles employ for electricity, and tiny house hookups mirror these. In a campground or an RV park, there is a means to connect your RV or camper to the electrical source via a cable. You can build a connection point for a cable like this into your tiny house. These cables come in various amp ratings, and those ratings are tied to how much electric use you anticipate in your tiny house. If you have a rather sizeable tiny house with a washer, dryer, dishwasher, air conditioning, electric heat, electric stove, and electric hot water, you're going to need a lot of electricity, and you will therefore need a higher amp rating on the cable. For a tiny house, you typically need a 30-amp or 50-amp rating.

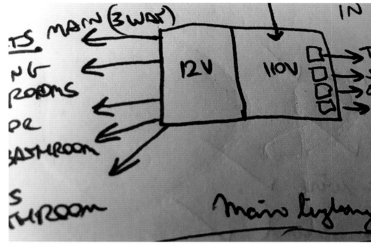

Planning out the electrical system based on the dual-purpose electrical panel I was using (12V/110V).

Once the electricity enters the house, it goes into an **electrical panel**. The concept of an electrical panel is in line with and very standard in a regular house. The difference in a tiny house is that the panel is going to be smaller because the energy needs are lower, and it will also be a mixed-use panel. When you plug into an exterior power source, it will come in as the standard 110-volt AC (alternating current). A mixed-use panel allows you to use 110-volt AC, but also has a

The power connector for the house is on the left. This happens to be a connector normally used on boats to get shore power. It features a 30-amp variety plug connection.

Powering It Up: Electricity

Low-consumption and voltage USB ports. We all have stuff we want to charge.

section where you can step down to 12-volt DC (direct current). The 110-volt AC runs larger appliances and supplies standard outlets. The much lower voltage 12-volt DC powers other various elements of your house that can run on 12 volts (see next paragraph). It's a good idea to have this setup because you may want to limit your electrical consumption and you may not always have access to an abundant and reliable electric supply. If, for example, you're in an environment where you can't plug the house into the electrical grid, you may be using a generator as an interim solution.

If you are running your house off of a **generator**, that generator is going to have a specific, limited capacity, and you may have an electrical shortfall unless you run as much as possible on the lower voltage. Typically people run their lights on 12 volts, plus have some USB ports for other low-voltage devices, like fans, and charging, like phones. There are many devices out there designed for the RV

community that run on 12 volts that make a whole lot of sense in a tiny house, too.

Any on-grid scenario means you are reliant on some form of external power source to fully power up your house. If you don't have a means of plugging the house in, well, there's not going to be much electrical activity in the house. That could be fine for you if you don't intend to ever be without an external power source, or if you utilize a generator. A large proportion of tiny house dwellers want something that's a little bit more independent, though, and they want to be able, at least for a short time, to go off-grid. So let's dive into off-grid systems.

Off-Grid Power

An off-grid scenario is one where there is no external utility company–supplied electricity needed of any kind. To generate your own power, you have three main options: solar, wind, and hydropower. Let's examine each one of these alternative sources in more detail.

SOLAR POWER

Solar power is provided and captured via **solar cells**, usually found in **solar panels**. These can be mounted on the roof of a tiny house, which is where many people place their array, or you can have it external to your house. The solar panels will be better located a short distance away from a shaded house in a place that gets as much sunlight as possible. Only when the sun is directly hitting the panels are you generating the most power, so even a partially shaded location is not ideal for your panels.

Solar panels provide immediate power that can be used for various utilities within the house. More importantly, however, you're going to want to capture that power and use it later. Storage of solar power is done using purpose-specific **batteries**. You can place a battery array into your tiny house and have it supply electricity even when the sun is no longer shining. This way you will still have power at night or on days when there is a lot of cloud cover. A solar setup without batteries doesn't make sense.

You can also use a **solar controller**, which determines the level of charge that you have on your batteries and then either adds more charge to depleted batteries or backs off when the batteries are fully charged. This monitoring prevents constant charging, which significantly diminishes the life of batteries. Nor do you want to constantly fully drain your batteries before recharging them. You always want your batteries to be at least 50% charged. In the end, your electrical usage and system should work together to ensure optimum resource use and battery longevity.

Batteries in an off-grid solar setup are **deep cycle batteries**, which are not the same as the battery you will find in your car. Car batteries are designed to

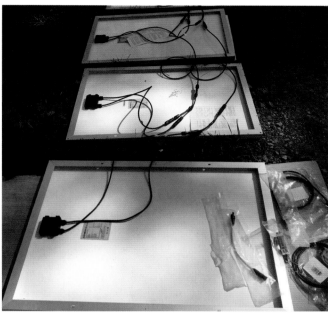
The backside of five solar panels. They are rated at 100W each; depending on your needs and the amount of sun that hits them throughout the day, that's quite a bit of power. They are all connected to one another.

A solar controller will help you keep your batteries charged to optimal levels.

Here are the five solar panels mounted on the roof of an off-grid house I built for a customer.

Powering It Up: Electricity

provide a lot of power for a short amount of time. They're effectively used to start the engine of the car, which takes a significant amount of power in a short burst. The power that you consume in a tiny house is not needed in short and powerful bursts, but rather as a low amount of power over an extended period. Imagine a reading light or a small desktop fan— they're not using a lot of energy, but the energy is being used over a longer time period.

The number of solar panels that you will need, as well as the number and capacity of batteries that you will need, are dependent on your location and on your power consumption needs.

There are also hybrid solutions available; it's not an all-or-nothing solar situation. You can mix solar power with wind power, or potentially hydropower, to get the power you need with the resources available to you. Read on for details about wind and hydropower.

WIND POWER

You can harness wind power using a miniature turbine much like the large-scale turbines you've doubtless seen dotting the landscape. Most people use wind as a supplement to solar, but you can also use it as a primary energy source. Perhaps your house will be located in a place where it is consistently windy; you could be on top of a slope that generally has strong winds or in a very open area like a field. Hybrid solar and wind controllers function in the same way as solar controllers. The wind turbine that you utilize to generate electricity will feed into the controller, which will charge or not charge batteries that are available to store that energy. Unlike solar panels, there is typically a brake circuit on a wind turbine. The controller will apply a brake to the turbine when no additional energy is needed. Applying this brake prevents the turbine from spinning constantly, therefore reducing wear and tear.

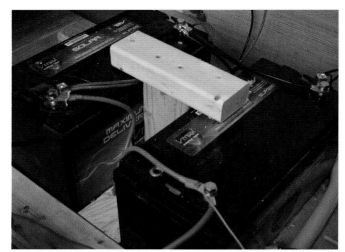

Two 12V deep-cycle batteries connected in parallel means that you are doubling the capacity, but not altering the voltage. It's still 12V— just a bigger 12V battery because there are two of them.

Power inverter: 12V in and 110V out. Power inversion has some inherent inefficiencies, but it's a great way to up-convert from DC to AC voltage for a variety of uses.

HYDROPOWER

Hydropower is a bit out of the norm for personal use, because generating power via water usually requires a larger operation. There are a number of hydroelectric power plants, like the one up at Niagara Falls, where the power is derived from the combination of gravity and water moving some form of turbine that then generates electricity. Hydropower, as it relates to tiny houses, is only an option in places where there is ready access to fast-flowing water. It's still relatively new technology, and not very commonly used for tiny houses, but keep it in mind just in case your tiny home is in a place where it is a viable option for you.

BATTERIES AND VOLTAGE

No matter what particular resource you tap for your off-grid energy, after the electricity is stored in the batteries, those batteries connect into a distribution system in your house. This system most commonly supplies the 12-volt DC elements of your house, since batteries commonly utilize 12 volts. You can hook the batteries up in ways to create higher voltages in a serial or parallel fashion, and you can also utilize 110-volt AC devices in your house through the use of an inverter. An inverter takes the DC voltage of the battery and converts it to 110-volt AC. This is not a perfect, lossless system: there is inherent inefficiency in converting electricity like this, so you can expect to lose from 10 percent to 30 percent of power in the conversion process. That said, an inverter is a great way to run larger household devices off of nothing other than solar or wind power. An inverter won't cost more than a few hundred dollars at most.

SUPPLEMENTING WITH PROPANE

Given the inefficiency of inverters and the fact that many appliances and AC devices utilize a large amount of power, there are specific scenarios in a completely off-grid setup that are generally not viable. For example, air conditioning units and electric stoves tend to use immense amounts of electricity. You can't design an off-grid setup to power them without including a prohibitive number of panels or batteries. You don't want to create a battery setup that's so large and cumbersome that it becomes too heavy and expensive to use to heat your house.

One good way to work around this is through supplementing with propane. Propane comes compressed in canisters like the tanks you see hooked up to gas grills. These canisters can be utilized in RVs and tiny houses. Often the tanks are mounted on the hitch end of the house because there's a built-in ledge where the house ends and the hitch assembly begins. Propane runs into the house and through the walls in piping much like electric and plumbing. Propane is terrific for heating, cooking, and generating hot water. There are other somewhat eclectic propane devices you can get, too, such as a propane fridge.

Propane needs replenishing eventually, and you'll have to monitor the tank pressure to make sure the system runs correctly. You're not entirely off-grid when you use propane, but many people opt for some form of a hybrid system that includes propane.

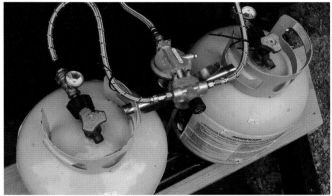

Propane tanks mounted on the hitch end of the house. Two are better than one and will last you twice as long as a single tank (obviously!).

A propane tank gauge lets you visually see how much propane is left in the tank. It's like the gas gauge in your car—you'll know when it's time to get more propane.

RUNNING WATER: PLUMBING

As with most other topics covered so far, tiny house plumbing is a little bit different than regular house plumbing. In regular houses, mostly PVC and copper are used for running hot and cold water lines. In tiny houses, a material known as **PEX** is generally preferred, and it's what I recommend. PEX is a very versatile synthetic plastic tubing material that comes in various diameters and is designed for various purposes. In-floor heating is a prevalent use of PEX in traditional homes. In the tiny house community, it's popular for plumbing for four main reasons: (1) it's useful in many different ways; (2) it's flexible; (3) it does not require any specific knowledge (i.e., soldering); and (4) it comes with very easy-to-use connectors. There is something called SharkBite® in the PEX plumbing realm that allows you to push connections together, and no further action or intervention is needed to create a watertight seal. I prefer the half-inch (1.5cm) variety of PEX. That thickness can deliver just the right amount of water, whereas some of the smaller diameter PEX configurations tend to be a little bit more constrained. The smaller the diameter of the tubing, the less water it's able to deliver. Water tank costs and fixtures aside, you can install the plumbing of an entire tiny house for a few hundred dollars—even less if you can borrow the crimping/specialty tools needed, since they can tend to be expensive and you may never have a need for them again.

PEX is a perfect way to run the water around your house. But how do you get water into your house in the first place? Well, similar to how the electric got into the house (described in the previous section), water systems work in an RV-style manner. A **water inlet** on the outside of the house allows you to hook a garden hose to the house. That garden hose water

PEX plumbing is ideal for tiny houses because it is easy to put in and great for all the tight spaces.

PEX lines branching off to supply the bathroom sink. Red is hot, white is cold. That just makes it easier to keep track of.

supply can be permanently hooked up and also provide the pressure internally in the house.

Alternatively, many tiny house folks opt for an off-grid variation for their water system. This means that there is a **water tank** in the house that gets filled up via the external connection. When you disconnect the house, the water system in the house is fed exclusively

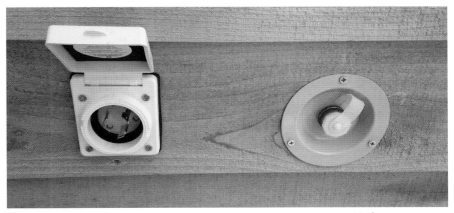

The water inlet is on the right. You can hook a standard garden hose up to this for your water supply or to fill freshwater tanks on the interior of the tiny house.

by the tank. This allows you to not have to rely on an external water source for either a constant supply of water or pressurization of the water.

Simply having a supply of water in the house does not create the pressurization you need to run the water, though. In conventional homes, the water is provided through pressurized lines, so you have to recreate that pressure in your tiny house using a **water pump** (which costs around $50 to $75). Somewhere near the water tank, you will install a pump that runs on 12-volt DC. When a tap opens, it creates a low-pressure situation, and there's a gravity-fed mechanism that feeds water into the pump. The pump senses that flow, kicks in, and pushes the water through the pipes and out of whatever fixture you've just tapped. This works for kitchen sinks, bathroom sinks, and the shower, and could also work for a toilet

A different type of water inlet. The principal and method is still the same.

The water pump and pressure tank are next to the water tank. Distribution of water to the rest of the house and the hot water heater branches out from here.

THE JOY OF TINY HOUSE LIVING

This freshwater tank is located under the bed. This happens to be a 46-gallon (175-liter) tank, which will supply water for quite some time if you take quick showers.

if you have a conventional toilet in your tiny house. A conventional toilet utilizes a significant amount of water and flushes like any toilet does in a normal house. We'll cover toilets in a later section (page 132), but most tiny houses do not utilize a standard flush toilet because, as you can imagine, they consume too much water. Much in the same way that energy and electricity need to be conserved in a tiny house, the same holds true for water. This is especially true if you have a limited supply from the water tank.

The internal water tank should be empty when you're **towing**; this is especially true if it's a large tank, because of the considerable weight of the water. We've all carried around a gallon milk jug (about 3.5 liters), so we know how heavy that is, but now imagine 40 gallons (151L), 50 gallons (190L), or even 70 gallons (265L) of water in a tiny house, and you can imagine how that creates a rather significant amount of weight.

Wastewater

If you're using water, you're creating wastewater. Where does it go? First of all, understand that there are two types of wastewater: gray water and black water. If you are washing your hands or taking a shower, the water that goes down the drain from those activities is considered **gray water**. This is not particularly dirty or polluted water. It might have some soap in it, but it's not a biohazard. Gray water is not such a big deal, and people come up with very ingenious ways to eliminate gray water outside their tiny houses. For example, you can create small troughs that the water runs through to water garden plants, assuming that you're careful about the amount and type of soaps and detergents that you add to the water. There are a lot of environmentally friendly soaps and detergents that you can utilize. Be aware that there are local restrictions surrounding gray water usage that may impact you, so check with your municipality. Gray water usage is an excellent way to recycle the water that you use in the house and give it a fulfilling second purpose instead of just letting it seep into the ground or evaporate.

Black water is what is created in a flush toilet; it is water that either has fecal matter in it or is in some other way not suitable to be drained or reprocessed without a lot more intervention. Black water is a lot more difficult to get rid of. You can't just drain it out of the house and onto the ground. One option is to have a black water holding tank and get it pumped out regularly; some people do this, but it's a rather expensive option, both in setup and pumping costs. Also, remember that flush toilets need to be supplied with water—are you going to have enough water to have a conventional toilet? For these reasons, unless you're in an RV park set up for the disposal of black water, most people won't opt for a standard toilet. So what do you use instead? See the section on toilets, page 132, for more information.

Hot Water

We all like to take showers, and hot water is generally something that we like to have when we take a shower. The best way I've found of heating water in a tiny house is with propane. In my first tiny house, I installed an **electric water heater**, and it works fine. I'm hooked up to the grid in that house, so I didn't have to worry about the amount of electricity used by an electric water heater, but I did have to worry about space. Most water heaters in conventional houses are rather large; they can take up half a closet. In a tiny house, space is at a premium, so that's not an option. I went with a very small water heater that fit under the bed in the second bedroom of my house, and it has a total capacity of 2.5 gallons (9.5 liters). I believed that I would be able to shower in 2.5 gallons (9.5 liters) of hot water because when you mix it with the cold water, you're getting 4 or 5 gallons (15 to 19 liters) total. Unfortunately, I discovered that I would invariably run out of hot water before I was done showering. So although the electric water heater does its job, in the limited space of a tiny house, it doesn't create an ideal shower situation for everyone.

When you have a **propane water heater**, the system works differently. The water heater, much like the water pump discussed earlier, senses a flow of water. It triggers a mechanism in the heater that lights a burner, fueled by propane, and that burner heats a chamber that the water flows through. As the water flows through, the temperature of the water rises, and then the water shows up at your faucet or showerhead, nice and warm. Propane water heaters are great

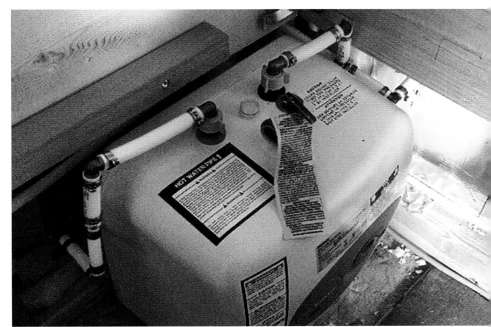

The 2.5 gallon (9.5 liter) electric water heater that I believed would be sufficient for taking a nice shower. It turned out to be a bit too small for the task.

because they provide hot water on demand; there's no ramp-up time. If I empty out my electric water heater and it refills itself with cold water, I have to wait for 10 or 15 minutes before that water reaches a temperature that I would want to shower in. With propane, you don't need to do that; water flows in cold and comes out hot. As long as you don't run out of propane (or water), you will have a steady supply of hot water.

STAYING WARM AND COOL: INSULATION

Insulation is an interesting topic, and one that's pretty important, although it's hidden in the walls. Insulating a tiny house serves a couple of purposes. I want to walk you through some of the things that are up for consideration when you choose how you're going to keep your place warm or cool.

Insulation is essential for two reasons. Primarily, insulation means **temperature control**. Insulation traps air in small pockets, which reduces the transfer of heat or cold. People often view insulation as a means of keeping something warm, but it's just as effective at keeping something cold. The fridge that you have in your kitchen is insulated—the exterior wall and the interior wall have some buffer intermediary between them to keep the cold in and to keep the heat out. A house is no different; it's merely on a bigger scale. You could be in a hot environment where you want to keep the house cool; insulation helps you do that. You could be in a cold environment where you want to keep the house warm; insulation helps you do that, too. A secondary effect of insulation is that it can **reduce noise**. If you are in a location where there's a lot of traffic or some other kind of noise, insulation will act as a buffer.

Remember that insulation also goes in the roof, and that other parts of your house are also part of the insulation equation: the door, the windows. I recommend buying a **door** that is fully insulated, and you will want to make sure your windows are at least **double-pane** glass (see page 68).

Let me tell you about part of my tiny house. I installed an antique brass porthole from a sailing ship in the main door of my house. It's very heavy and very old, and the glass is single pane. In the winter, I get serious thermal bridging through this window.

Insulation. You can't see it, but it's in there, and it's reducing not only heat transfer but also noise.

If you touch the metal on the inside of the door, it is bitterly cold—the cold from the outside passes right through the metal. Then, when you couple that with the single-pane glass, the warmth inside the house is meeting up with the cold outside the house, and the result is condensation on the warm interior side. Even though it has sometimes been so cold that the condensation freezes on the inside of the glass, this isn't the most terrible thing. However, it's definitely not an effectively insulated spot. Thankfully, it's just a small window, only about 8" (20cm) wide and 13" (33cm) tall, so it doesn't make my house cold. But it does go to show what happens when you don't correctly insulate everything.

There is also the notion of having too much of a good thing, and this applies to insulation. You don't want to completely cocoon yourself in an **over-insulated space**. If you close everything off too much, you'll get a lack of airflow, which can cause condensation, mold buildup, carbon dioxide buildup, and other adverse consequences. A house still has to breathe, so you have to build in ventilation. You can add passive air inlets and ventilation fan systems to help keep the environment balanced (more about these on page 127).

Not an exciting picture, but this is a what a passive air inlet looks like. This one is located under a bed to minimize any chance of feeling a draft in the living space.

This is the door on my tiny house with the antique ship porthole. It's solid brass with single-pane glass, so it looks great, but it's one of the places where thermal bridging readily occurs.

Types of Insulation Materials

As mentioned earlier in the book, R-value is a representation of how well insulated your house is. Insulation efficiency is just a measure of how well a material traps air and how (in)efficient it is in transferring heat or cold. Different insulating materials have different R-values, the higher the better. The specific R-values you are able to achieve with your chosen product will be clearly listed on the product itself, and depends in part on thickness/quantity: a thin layer of a super-efficient insulation may have the same end R-value result as a thick layer of a less-efficient insulation. Also consider ease of application. Spray foam insulation may be the most expensive, but it's also the quickest to achieve a relatively high R-value. You cannot rank the types of insulation neatly in order from lowest to highest R-value because different variants of one product will have higher or lower R-values than different variants of another product. Consult the exact products you are comparing.

Pink glass wool settles, and so should only be used in non-mobile tiny homes.

PINK GLASS WOOL: This is the most popular form of insulation in traditional houses. You've seen it if you've been to a hardware or home improvement store and wandered down the insulation aisle. It is spun out of glass, and the production process is similar to the way that cotton candy is made. Pink glass wool is very inexpensive, which is a big driver when it comes to building full-size houses. It is designed for a stationary environment—to go into a wall and just sit there. It's not designed to be jostled, moved around, or encounter any of the special stresses that it would encounter if you used it as insulation for a mobile tiny house. Can you use it in a tiny house? Yes, absolutely. But here's what's likely to happen as you're driving down the road with your house. The bumps and road irregularities that you subject your tiny house to cause vibration, and that vibration causes the pink glass wool to settle, or compact, within the walls. At some point you'll be left with a completely uninsulated section at the top of the wall. For this reason, I don't recommend pink glass wool for mobile tiny homes. If you are installing this kind of insulation, you will need to wear gloves and some form of face protection, since the fibers can cause irritation to the skin and airways.

COMPRESSED FOAM BOARD:

When I built my first tiny house, I opted for compressed foam board for insulation. This is a rigid foam that you can buy in different thicknesses. It comes in 4' x 8' (1.2 x 2.4m) sheets, just like plywood does. Imagine the framing of the house with cavities within the wall between the 2x4s (50x100mm). What you do with foam board is cut out pieces that roughly fit into those cavities, leaving a bit of a gap around the edges, until all the

You have to cut foam boards (these have a metallic liner) to size to fit into the various framing cavities. This takes a long time.

Another house insulated with foam board.

cavities in the house are filled. It's a somewhat labor-intensive activity; no two spaces in the wall are going to be identical, so it involves a lot of cutting. The foam board is pretty easy to cut through, though; you can use an extended blade box cutter/utility knife to do the work. Foam board is an okay way to go with insulation, and I'm still happy that I went that route with my first house. To cover the gaps around the exterior of the foam board, use an expanding foam material that comes in a can, sprayed along the edges. This locks the board into place and provides a little extra insulation around the edges so that you don't get drafts.

Rock wool is my go-to insulation material.

ROCK WOOL: More recently, I've moved over to using primarily rock wool for insulation; my preferred brand is offered by a company called Roxul. Rock wool is somewhat similar to pink glass wool, but it is a lot denser. It has the consistency of a loaf of bread; you literally can cut it with a bread knife. This material fills gaps and spaces very effectively and is readily available. It's also water resistant; if you test it by drizzling some water on it, you'll see that it beads right off. Rock wool is also fire resistant, because it's made of rocks. For all these reasons, it's a rather good choice for tiny house insulation. The product itself comes in mats of different thicknesses, including thicknesses designed to insulate the spaces created by 2x4 (50x100mm) framing. Make sure that you cut the material to a greater width and height than you need to fill in a space—you want a bit of compression on either side of the material to hold it in place. Like pink glass wool, rock wool is an irritant, so protect your skin and respiratory system while working with it.

NATURAL MATERIALS: Some people want to go with a more natural material for their insulation. For example, there is an insulation material made from old, chopped-up denim. It is prone to settling, like pink glass wool. It is not irritating, though. There are two other issues to consider with a natural insulation like this. First, moisture can be a significant problem, leading to mold and rot. Second, these materials are quite vermin-friendly; mice or other pests would be thrilled to turn your insulation into bedding, which is not something you want happening in your walls.

SPRAY-IN INSULATION: There are a lot of different types of this material. No matter the particular type, a company comes in with a bunch of equipment and sprays foam into all the cavities of your house. The foam then hardens during a curing time. Having your insulation done this way is quick and easy; you go from an uninsulated house to a fully insulated house a single day. However, you'll pay a premium for the luxury of not having to do all that work yourself. Once you've decided to go with spray-in insulation, you will need to make sure that you are ready for the installers when they arrive: your wiring and plumbing must be completely done, and you must be ready to close up the walls once the insulation is in. You can also use spray-in insulation on the bottom of your trailer.

Spray-in insulation must be handled by professionals.

COVERING IT UP: ROOFING

The roofing material you choose for your tiny house is important. The roof of the tiny house will not only be exposed to the elements, but also to a lot of wind and harsh forces when traveling down the highway. Not all roofing materials are suitable for tiny houses. When you think of traditional houses, you will most often see an asphalt shingle roof. This is a roughly textured roof made of petroleum products and some granular rock materials. These are perfectly fine for traditional houses because they tend to be rather heavy and are not designed to withstand hurricane-force winds. If a hurricane does hit, it's most often the roof that sustains some form of damage. In a tiny house, you don't have the luxury of being able to use heavy materials, nor can you afford to choose a material that can't stand up to extreme wind and weather forces.

Tiny houses can have various types of roof, but by far the most popular is **metal**, and that's what we'll focus on here. Using metal works well on tiny houses for many reasons. For one, it's lighter than traditional materials (such as asphalt shingles). A metal roof is also much stronger and has a higher resistance to wind forces if installed properly with the manufacturer-recommended number and type of fasteners. Metal roofing can be painted any number of different colors. Sheets of metal are produced in both steel and aluminum in various thicknesses (gauges), with aluminum tending to be lighter weight. These metal sheets often are interconnected by something called a **standing seam**. This is achieved using flat or semi-corrugated pieces of metal that come in various widths. Very often you'll see these in 16" (40.5cm) widths, which is the standard on-center spacing for roofing rafters. At the edge of these sheets of metal, there's a flap that folds up, and on the next sheet, you'll find a groove that slots on top of that raised flap. This creates sheets of metal that are fastened to each other and fastened to the rafters (which are the lumber components of the roof framing that lie parallel to each other under the roof sheathing). This forms a very tight weather- and wind-resistant roof on a tiny house. You will have to expend a little extra effort to obtain small quantities

Here is an example of a panel that has the raised edges, where one panel meets the next.

Multiple roofing panels interlocked with one another create a standing seam metal roof.

of standing seam material, because many of the companies that produce this particular material are geared for more substantial sales of larger roofs.

Metal roofs have a couple of downsides. First, they're **more expensive** than standard roofing materials. A shingle roof is around $1 per square foot (material cost), but shingle is entirely unsuitable for tiny houses, as we've seen. Metal roofing—depending on the quality, thickness, and color—is much more expensive, easily five times more per square foot. Although this is a consideration, it's typically not a huge deal, because you're not covering a large area on top of your tiny house. Therefore, the increase in price is not that significant. Another worry that people have about metal roofs is that they're **noisy**. You can imagine rain hitting a metal roof. Yes, they can be noisy, but the noise can easily be combatted and drowned out through effective use of insulation in the roofing members. Will you still hear a metal roof a little bit more than you would a conventional asphalt shingle roof? Yes, absolutely. However, the tradeoffs regarding safety and longevity more than make up for the slightly higher noise level that you will experience with a metal roof.

Roof Design

Before any of the roofing material goes on, you'll have to ascertain the roof design you want, which includes determining the pitch and slope of the roof. You never want to design any house with a flat roof—that's just an invitation for water, snow, and leaves to gather with no way to slide off. Snow and ice in particular can be incredibly dense, and if you don't have some means for them to come off the roof, you

Metal roof in detail. You can see how this is a sturdy and good way to go on a tiny house.

open yourself up to a worst-case scenario where your roof caves in. Water pooling and standing on top of the roof can also lead to leaking and other problems over time.

You have a couple of basic types of sloped roofs to choose from: a gable roof and a shed roof. These vary regarding their pitch and construction. A **gable roof** is where two sides of the roof rise from the edges and meet to form a peak at

GABLE ROOF

SHED ROOF

Extending the roof edge past the side of the house keeps water off the exterior walls of the structure.

the top. You can have more complicated gable roofs, but that is their essence. A **shed roof** is one that doesn't pitch up from either side, but instead has just a high side and a low side—a single slope instead of two slopes that meet. A shed roof is easier to build because you're effectively just building a box, then lifting one edge of the roof up to create the slope. The lower edge of the roof can rest on the framing box that you've created.

At a minimum, you should build a roof with a **slope** of about 1" (2.5cm) to 1½" (3.8cm) drop per foot (30cm). This can go all the way up to 2½" (6.3cm) or

3" (7.6cm) per foot (30cm). But if you get too extreme, you're likely wasting space that you could be using for your living area.

A roof should not simply terminate at the edge of the house; you want a little bit of an **overhang**. When water runs off the edge of a roof, you don't want it running down the side of the house. If you extend the roof edge out, the water will drop from there. Most traditional houses have gutters on the edge of their roofs, but most tiny houses do not, so that little bit of overhang is critical for protecting your house.

Flat roofs are a no-no: snow and water need to be able to slide off a roof easily.

BEAUTIFUL INSIDE AND OUT: FINISHING TOUCHES

You've done all the framing and the building, or maybe it's been done for you. Now you're left with finishing the interior. This is the fun part, your chance to be a decorator. The inside of a tiny house can make or break the house. There's nothing worse than a beautiful exterior that hides a bleak interior. Following are some suggestions to give your home a custom look and a real personality.

FAKE IT: Not everything has to be functional, of course—even if it looks like it is. Add interesting details that break up empty spaces, like the faux ceiling beams shown here.

This is a rather bland, boring ceiling.

Some simple stained cedar slats add a lot to the overall feel of the interior.

FOLLOW YOUR INSTINCTS: Have the interior be a reflection of you. It doesn't matter whether you like modern things or traditional things or eclectic things— go with what you like.

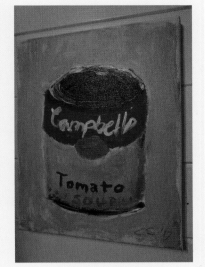

I'm no artist, but that doesn't mean that I don't have the right to pick up a paintbrush. This painting may not be great, but it's intensely personal and meaningful to me, since it's my artwork.

KEEP IT SIMPLE: As a general rule, too much or too busy doesn't work in a tiny house. Look for inspiration in places like Pinterest or other websites to see what you can do in a small space and have it look good. And of course check out the gallery on page 10.

An uncluttered, clean look in a gypsy wagon build I did. Less can be more. Good use of a small space creates a warm, not sterile, feeling.

AVOID CLUTTER: Clutter in a tiny house makes it feel claustrophobic and smaller than it is. Put everything in its place, and create spaces to store items out of sight.

Everything kitchen-related that I keep out and visible has to be appealing. Think about what you use. Paper plates would not look good on these antique shelves.

STICK WITH A GENERAL THEME: Don't try to mix a lot of different styles or genres of decorating in one place. If you're going for a traditional look, then stick with that. If you're going for a modern, clean look, then stick with that.

This interior has a consistent, modern style with a bit of flair; it doesn't veer off in multiple directions.

USE QUALITY MATERIALS: Tiny houses are not about producing something as cheaply as possible. If money is tight, then there's nothing wrong with decorating the interior of your house with good-quality things that are repurposed.

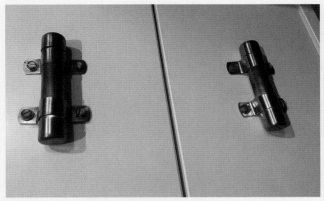

I love working with copper. It ages nicely and has a warmth that other materials don't. I made these drawer pulls out of standard copper plumbing hardware.

Untreated cedar on my porch. It weathers so nicely, and I really like the gray. You could stain and seal it to prevent this from happening, but I chose not to.

THINK ABOUT COLOR: Colors are really powerful inside a tiny house. If you go darker, do so very selectively, because dark colors can make a space feel smaller. In my tiny house, I painted all the walls white, then went with a dark ceiling because I thought it looked good. I also have recessed lighting in the ceiling, and that lighting pops against the deep blue color of the ceiling.

A dark pop of color on the ceiling works great with the white walls.

MAKE SMART LIGHTING CHOICES: This is a very crucial aspect to your tiny home. We've already gone through why choice and placement of windows is important in a tiny house (page 67). The artificial lighting that you use in your tiny house is also of great consequence, so choose carefully. As I mentioned earlier, I'm a fan of recessed lighting, but I'm also a big fan of filament Edison bulbs because they create a real mood. I use those rather liberally in my structures. Skateboard lamps are something that I've designed, which do a nice job of showcasing a used item that is infused with a lot of character. Also consider indirect lighting, which is lighting where you don't see the actual light source because it's covered or masked in some way.

I make these skateboard lamps; they are a great way to add character and light to a space at the same time.

In the bathroom of my tiny house, I set the mirror slightly off the wall and did the lighting for the bathroom behind the mirror; you don't see the lights themselves, just the light emitting from behind the mirror and creating a nice effect. Incidentally, the mirror in my tiny house is something that I bought at an estate sale. It's a repurposed antique piece that was not only a financial steal, but also tied in well with a lot of the other finishing touches that I used in my house.

Edison bulbs are very retro. They evoke a coziness that no ordinary light bulb can.

Placing LEDs behind an object for indirect lighting. This is the back of a bathroom mirror.

The mirror in the bathroom of my tiny house. Indirect lighting emanates from behind the mirror. Good lighting is so important for a tiny space.

THE JOY OF TINY HOUSE LIVING

INCLUDE SIGNATURE, STANDOUT PIECES: A fun way to jazz up the interior of a tiny house is to include pieces that are very distinctive, possibly pieces that wouldn't really make sense in a regular house. For example, I had to install a ladder to the loft sleeping area in my tiny house. I could have just cobbled something together from a bunch of 2x4s (50x100mm), but instead, I spent a little bit of money on an antique library ladder. This is the type of ladder that rolls back and forth on a bar, in my case on the edge of my loft area. This ladder not only gets me up into the loft, but it's also a visually appealing centerpiece to the living area. Plus, it's easy to move, tucking into the corner on the edge of my kitchen and thereby saving space.

Another item that I repurposed in my tiny house is what turned into the kitchen. I could have bought some kitchen cabinets and created a kitchen that way. I could also have constructed some cabinets myself. Instead, I opted to buy a sideboard buffet that you would have in a traditional dining room. I turned that into my kitchen counter, but I also did cutouts into the surface of it for a drop-in sink and a drop-in cooktop. That piece of furniture also had the advantage of being large enough to house a dorm room–style fridge on one side of it. The fridge is incorporated into the kitchen area, but isn't visible, which is nice because dorm room fridges don't look that attractive.

The library loft ladder in my tiny house rolls out of the way to the left and between the kitchen and wall. This way, it doesn't take up any space, but still looks great.

My kitchen. Formerly, it was a piece of furniture meant for a traditional house.

Gorgeous detail on the wheels of the library ladder. This is a highly functional and beautiful historical piece.

My dorm-room fridge is conveniently hidden in my kitchen furniture creation.

Beautiful Inside and Out: Finishing Touches

STAY PROPORTIONAL: Proportions are important in a tiny house. Overstuffed easy chairs and big couches tend not to work in a small space; you invariably make the space look even smaller than it is by introducing an oversized piece of furniture into it. For my tiny house I purchased a rocking chair that is very much on the small side, much smaller than a traditional rocking chair. However, it is still big enough for an adult, and it looks very much at home in my house because it conforms to the sizing of the rest of the elements in the house.

Anywhere else, this rocking chair would seem to be on the small side. On my tiny house porch, it looks just right. It's actually quite comfortable, even for a big guy like myself.

ANDREW BENNETT

Andrew: I'm Andrew Bennett; I own Trekker Trailers and have been building tiny houses and custom micro campers for eight years now. I've built over 150 so far. I specialize in gypsy wagons, and now I'm hitting the concept of providing an affordable tiny house solution.

Chris: *So you've been doing this a long time, and you've clearly built many houses. What have you seen change over the years doing this?*

Andrew: The industry and the business of it. I see a lot of people are trying to figure out how to make a career or life out of it. I think we're seeing a little bit of a paradigm shift in this country; keeping up with the Joneses is now, "Hey, let's just be smart."

Andrew Bennett of Trekker Trailers with his son.

Chris: *What impact has that had on the overall tiny house movement?*

Andrew: I think it's sparked healthy competition; more so than that, I think it's snowballing the creativity. For a while, every tiny house looked the same. They all had a loft, and they all had the cute roofline, and a front porch. Now you see so many different varieties of styles and building processes and practices.

One of Andrew Bennett's many tiny house creations.

Interview with ANDREW BENNETT

Another of Andrew's creations; this is a vending model.

Chris: *Take your crystal ball and look a year or two into the future; what does that look like?*

Andrew: I see it affecting our government. Rather than trying to stick with the status quo when all this stuff's going on politically right now, people are looking for something different. This is different. It's not the norm, but there's something good to be had from it.

Chris: *It is true that a lot of people try to categorize tiny houses as a fad, a trend, a movement. Where on the spectrum does it fit for you?*

Andrew: Well, people who say that about something tend to later realize, "Hey, this is pretty useful," and then eventually, "Hey, I can't live without this." That's how it went with cell phones. How many people thought that cell phones would be around forever when they were first in a big bulky bag sitting in the center console of your '85 Mustang? And now our cell phones are more technically capable than the NASA stuff back then.

JOE AND KAIT RUSSO

Joe: Back in 2014, we were looking for a way out of our corporate careers. We lived in Los Angeles and had a house. Kait came up with the crazy idea for us to quit our jobs, sell everything, and buy an RV and hit the road. We were only going to do it for a year, but once we got on the road for a month, the bug hit us; we haven't been off the road since. We've been on the road about three years now. In 2017, we went from a 30' (9m) motorhome down to a just-under 21' (6.4m) van, and for us, it's been working out well.

Joe and Kait Russo of *www.weretherussos.com*.

Kait: I think the biggest thing for us making that transition from life before to what we live now is changing our way of looking at life. I had this American dream: you work hard to climb the corporate ladder, you buy a home, settle down, have kids—all of that stuff.

Joe: ...and fill it with stuff.

Kait: Yes, and fill it with stuff. And then my head just came out of the clouds one day, and I started questioning, why am I doing these things, and who says I have to do these things? I realized that life wasn't fulfilling or making me happy, and so I sought out something that would do those things. That transitioned into living in a tiny home on wheels and traveling, working for ourselves and connecting with like-minded people, and inspiring others to show them that there's a different way to live.

Chris: *What made you choose the van option for yourselves instead of a traditional tiny house?*

Joe: We wanted something that fit into a regular parking spot. A van was a natural answer for that, and we liked being wholly contained. For us, the van made the most sense.

Kait: We chose this tiny home on wheels for its freedom and flexibility: the freedom to roam and go wherever we want, and the flexibility to go off grid or hook up at a campground and still be able to work for ourselves and live our dream—which is to travel.

Joe: Right. We wanted to enjoy the travel aspect, but at the same time, take our house with us.

Kait: Exactly. The example we always give is that we love national parks, and sometimes the scenic pullouts are very limited regarding how big you can be to stop there. There are times where, if we're going on the Rim Drive at Crater Lake and there's a little scenic pullout, we can stop, open our door, have lunch, take a nap, or work for the day. That national park is our backyard, and having a bigger structure, like when we had our 30' (9m) motorhome, meant those scenic pullouts weren't viable options for us to stop at.

Chris: *You referenced earlier corporate jobs, and now you're referencing working from the road. Is it more freelance type work that you're doing as you travel the country, or what does that look like?*

Joe: We had to figure out a way to be on the road for longer than a year. So we started a website, *www.weretherussos.com*, and also a YouTube channel, and we started creating content. As part of that strategy, I wrote the book *Take Risks*. That keeps us on the road and puts gas in the fuel tank, and we now work for ourselves.

Kait: Part of the primary goal of the content that we're creating is to help share the knowledge that we've learned through this lifestyle in order to help other people either transition to similar things to what we're doing or do something completely different.

A full awning is built in on the slide door side.

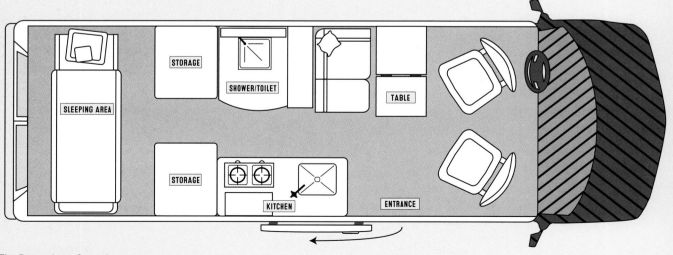

The Russos' van floor plan.

THE JOY OF TINY HOUSE LIVING

The Russos' Hymer Aktiv 2.0. It's a Ram ProMaster Conversion Van.

Chris: *What are the downsides? Are there any regrets that you've had since you did this?*

Kait: Our number one regret is always that we wish we had done it sooner.

Joe: Regarding downsides, when we first got in a tiny structure, it felt claustrophobic. Kait and I had a tough time learning how to use the space together. But as we've spent more time in it, this has become home; it now feels spacious to us. When we go into a tiny house, we're like, "Man, this is a mansion!" It's not for everybody, but once we made that mental switch that this is home and learned to live and found our groove in the space, all the things that we might have seen as a disadvantage or difficult have just become part of our daily life.

Chris: *So one of your regrets is not doing this sooner. That implies that you have no intention of stopping this anytime soon, either.*

Joe: No, we've found our happiness, and that is just working for ourselves and being our own bosses, and being able to spend as much time as we want around each other. For us, that is everything we want, and whether we do it in a van with backpacks or find a ranch somewhere and throw a tiny structure on that, it doesn't matter where it is, as long as we're doing it together the way we want to do it.

OTHER CONSIDERATIONS

Spending time in a tiny house has its own particular nuances. A tiny house on wheels is not like anything you are accustomed to. There will be things that you are not used to that will need a bit more thought and attention than they would in a traditional house or apartment. They are neither good nor bad, but do require some adjustment on your part.

NOT YOUR NORMAL KITCHEN: COOKING

Cooking in a tiny house has several unique considerations to keep in mind.

SPACE: You have a lot less space to cook in, period. Counter space, space to turn around in, space for more than one person to cook at a time, space to take something hot off the stove. You'll have to be prepared for this; otherwise, you could accidentally burn yourself or a surface or make a mess that is not easily cleaned up.

Yes, that's a small kitchen. This is from my gypsy wagon build. If you want to cook, you have to bring out a portable induction cooktop. Why have that take up space when not in use?

ODORS: When you're cooking in a small space, you're going to very quickly fill the entire house with the scent of whatever you are cooking. This could be a good or a bad thing. Many people like the scent of bacon, so having that everywhere might not be so bad. If you're cooking fish or a cheesy dish, though, it might not be so great. Take steps to ventilate or choose what you cook wisely.

You may want to leave the shrimp recipes for someone (or somewhere) else.

REFRIGERATOR: The fridge will be smaller than you are used to. You may also have to go without a freezer. This can be a good thing, though, as it forces you to consume primarily fresh food.

Small 12V fridge. Works great using solar power in an off-grid house I recently built.

Cooking with a limited fridge more often than not means cooking with fresh ingredients.

STOVETOP CAPACITY: You may only have a two-burner cooktop; therefore, you won't be able to have four pots going at the same time. That doesn't mean you can't cook a four-course meal, but it does mean you will have to plan your cooking process out in more detail than you're used to.

Two-burner propane cooktop.

TYPE OF COOKING SURFACE: The type of cooking surface you'll be using may vary from what you're used to. Induction cooking is a popular option in tiny houses, as is gas/propane. Gas, of course, creates additional heat in the tiny house. Induction cooking is as popular as it is because the actual cooking and heating surface does not heat up—only the food in the cookware heats up. It is a good option if you do not wish to create additional heat in the house.

Induction cooktop. I opted for a single burner, since I cook much of my food outside on the fire pit. It's great for cooking a pot of pasta or heating up the coffee water in the AM.

SLOW COOKERS: Though it may seem like an ideal solution in a small kitchen, you are not going to want to use a slow cooker in a tiny house. Slow cookers produce a lot of vapor, which translates into humidity in your small structure. In a typical, regular-sized house, this is not a big deal because the humidity dissipates, but in a tiny house, it can become a real problem. (Read more about humidity on page 127.)

See all that condensation? You don't want that filling your tiny house.

OVEN: Not many tiny houses have an oven. Cooking a turkey or baking a cake—or even popping in a frozen pizza—may no longer be an option for you. You can use a countertop toaster oven if you must—but remember, that will hog up counter space. Everything is a tradeoff!

A toaster oven is likely the closest you'll get to a real oven in a tiny house.

The bottom line is that doing the gourmet thing in a tiny house is harder. It's not impossible; it just requires more thought and more of an organized process. You're not going to have things like a hooded vent fan over your stove, and you're also unlikely to have all the various implements that you might have in a regular house. There's very little room for a blender or a food processor or a food dehydrator. As is always the case, tiny houses are a collection of tradeoffs, and if cooking is important to you, you will want to dedicate some additional space to your kitchen.

A lovely tiny house meal—no cooking required.

VENTING SMALL SPACES: HUMIDITY

Humidity is not your friend in a tiny house. Short-term humidity is not an issue, but if you create an environment where you constantly have high levels of humidity, you're effectively inviting **mold** to live with you. A poorly ventilated tiny house can become unlivable pretty quickly. Just being in a small space—breathing, showering, and cooking—is going to raise the humidity levels in a tiny house. Let's discuss some ways that you can make your tiny house breathe and, by extension, keep the environment healthy.

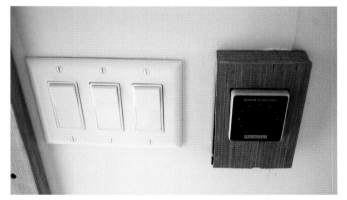
I placed this vent speed controller next to light switches. Dial in how much venting you need at any given time.

Dissipating humidity is all about **venting** the tiny house properly. You may be fortunate enough to have your house in a location where you can keep your windows open most of the time to create cross-ventilation. This may effectively mitigate any humidity issues you have. But even in the best of environments, you are going to need to close the windows and seal the house off at least some of the time. That's when you will face a humidity and condensation issue. You need some form of air exchange in the house; you need to push stale, humid air out of the interior and take in fresh external air.

The ugly truth is that humidity is more frequent in tiny houses, and humidity encourages mold.

Venting can be as simple as a fan mounted in the wall that pulls or pushes air in or out of the house. If you push air out of a house via a vent, air also needs to be able to get into the house to take the vented air's place. A **passive air inlet** is the way to go to solve this issue. The inlet is passive in the sense that it just sits there and there are no moving parts. It simply provides a way for air to get in to replace the stale, humid air you are pushing out. Pulling air in can be undesirable, particularly in very cold environments, so you need to be cognizant of how and where you have that air enter the house.

Based on your activities in the house, you will have higher or lower venting needs. If you're cooking or taking a warm shower, you're going to create more humidity for a short period that needs to be dealt with. You want to address different humidity levels in different ways. Vent fans that are on a **speed controller** are ideal. In one house that I recently built, I installed what's called a **bilge fan**. These are used on boats to extract air from the engine compartment, which, in a boat, can be dangerous, because gas fumes can gather and ignite and cause an explosion. Bilge fans are designed to draw that air

out quickly and clear out a small space. These fans are not big and they're easy to install, but they can be noisy, and therefore hooking one of these up to a speed controller is a great way to be able to tone down the noise and dial in the specific needs that you have for venting.

Another means of cutting down your humidity is a **mini-split**, which is a special type of small heating/cooling unit. If you live in a warm, humid environment, such as Florida, this is a good solution. Air conditioners limit moisture and condensation, and by using one, you pull moisture out of the air. This is part of the reason why air-conditioned rooms feel more comfortable—not only are they blowing cold air into the room, but they are also removing some of the humidity from the room.

HEATING AND AIR CONDITIONING SOLUTIONS: MINI-SPLITS

Rustic wood-burning stoves are a popular heating option in a cold environment, but in a structure that is intended to move around, they are a bit too heavy duty. More versatility is usually the goal, and in general, having a way of both heating and cooling a tiny house efficiently and easily is a worthy goal to strive for. Some homes have the dedicated **window air conditioning units** that most people are familiar with. Heat is then provided via **space heaters** that plug into standard outlets in the house and that can be packed away when not needed. This two-prong scenario is indeed cost-effective and straightforward. All it requires is moveable/removable units that use electrical outlets. But the most popular choice for medium to large tiny houses is to utilize what's known as a mini-split.

A **mini-split** is a ductless heating and air conditioning unit in one. Let me first go over what a mini-split is and how it differs from traditional air conditioning units. A window-based air conditioning unit is one that sits in the wall or the sash of a window. It is a self-contained unit that pushes or creates cold air for the interior of the house, and there's a heat exchange that takes place on the exterior. If you've ever put your hand up to the outside of one of those window or wall unit air conditioners, you've felt the heat being pushed out of the unit in the back, while on the inside, the unit is blowing cold air. A unit like this is self-contained: all of the components reside in that single metal box. A mini-split, on the other hand, is a split unit, as the name implies, so, in other words, part of the air conditioner resides on the interior of the house, and part of the air conditioner resides on the exterior of the house.

A wood-burning stove can look either rustic or modern, but ultimately may not be the choice for you.

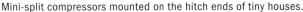
Mini-split compressors mounted on the hitch ends of tiny houses.

The exterior piece is the **compressor unit**. The compressor unit compresses down the refrigerant. That refrigerant is then sent, compressed, into the inside of the house, where it arrives into the **wall-mounted interior unit**. That's where it's decompressed, creating the cooling effect. The key here is that the two units that make up a mini-split are separated, connected by a set of hoses. Why is this so good for tiny houses? The compressor unit often resides very neatly and nicely on the hitch structure of the tiny house, so that's a popular, out-of-the-way location for these units. They can also occupy space higher up on the roof line or a gable of a house; wherever they go, they can be **reasonably unobtrusive**.

From wherever the compressor unit sits, the hoses will lead to the interior. This is where the next advantage is. **You don't need to cut a big hole** into the side of your tiny house for a mini-split. You only

need to cut a hole for those hoses. They're usually situated at a somewhat high point on a wall that has a suitably large enough area to house the unit. The size of the interior unit will vary, but they are typically large. They tend to be wider than they are tall, but, as I said before, they can be placed in an area that's pretty unobtrusive. Inside that interior unit, you have the coils that produce either the cold or the heat—whichever you need at the time—and you also have a blower unit so that it can pass air over those heating or cooling elements. This creates the heating or cooling effect within the house. The units generally look like white plastic boxes. This can be a little bit jarring if you have a very rustic or traditional interior in your tiny house. If you are really motivated, you can wrap the interior unit in another kind of box to suit your décor.

Another advantage to mini-splits is that they can be hooked up to a **standard thermostat**. This allows you to control the temperature within a tiny house in the same way that you would in a traditional house.

Yet another positive aspect of mini-splits is **maintenance**. These units are easy to maintain and repair because of their accessibility. If there's an issue with the interior unit, you deal with the interior unit, and if there's an issue with the exterior unit, you can deal with that. In both cases, they're readily accessible, because you're not integrating the units themselves into the structure of the wall.

Mini-splits run fairly **quietly**, and they're also **relatively inexpensive** for what they are. Sure, they're going to be more costly than a cheap window unit air conditioner, but the price points that they start at are not high (see below).

Efficiency is yet another plus. Mini-splits can even be run using solar-based systems in warm environments that have a lot of sun.

Ultimately, while it might not be as romantic as having a wood-burning stove in your tiny house, a mini-split is a worthwhile consideration when it comes to heating and cooling, so think about whether it may be right for you. Implement the method that best suits your environment and the climate conditions you are likely to face. There is no one-size-fits-all answer for heating and cooling. Categorizing your options by cost from least to most expensive looks like this:

Here is the inconspicuous internal component of the mini-split, pointed out with a not-so-inconspicuous red arrow. It's perfect for tiny house applications.

- **Space heater**............................. $35
- **Vent fan with passive inlet and speed controller**............... $150
- **Window unit air conditioner** $150
- **Wood-burning stove**................ $500 to $1,500
- **Mini split**................................ $1,500 to $2,000

WHAT EVERYONE IS CURIOUS ABOUT: TOILETS

Toilets are such an intriguing part of tiny houses. On my website that documented the build of my first tiny house (*www.tinyhouseinthecountry.com*), the blog post regarding toilets has been by far the most popular post on the entire blog. Why are people so intrigued by tiny house toilets? It's one of the aspects of tiny living that deviates the most from what we are used to in our Western, modern civilization. It's a marked departure from what we accept as normal. I can't think of a time I ever used anything other than a regular flush toilet in a traditional house. Plumbing and toilet conventions don't vary much.

But when you're living in a tiny house, you are most likely not going to have a **conventional flush toilet**. The location of your house and what services you have at your disposal play a crucial role here. You may have access to a public sewage hookup, which would allow you to have a traditional flush toilet. But since tiny houses are transportable, they often end up in locations where it's just not feasible or viable to have a regular flush toilet. Plus, flushing a regular toilet uses a ton of water. If you don't want to lock yourself into one kind of location, you'll need to make a different choice. As with all the other tradeoffs that we've discussed in this book, if a traditional toilet is important to you, you can make it happen, but more often than not you will have to utilize some other form of toilet. Here we'll go over several types of toilet options, along with their advantages and disadvantages.

Composting toilets ($500 to $1,500 depending on manufacturer and features) are the frontrunner in tiny house living, and they run an entire spectrum in and of themselves. In its most basic, and by extension least expensive, form, a composting toilet is nothing more than a 5-gallon (19L) bucket that is incorporated into a box and that has a traditional toilet seat on top. First you use the toilet, then you put a medium into the bucket. That medium could be something like sawdust; some people use the discarded remains

The Separett composting toilet with an externally vented chamber and urine separation. This is a high-end composting solution.

of the coffee-roasting process. There's any number of different things that you can put in there that will draw moisture out and diminish odors. It's not unlike the way a cat's litter box works. The compost that results from this process is a great fertilizer. However, it should not be used on edible plants.

People have come up with ingenious ways to make the simple more complicated, and therefore more expensive; this means you can buy **composting toilets with churning mechanisms** built into them. Nature's Head is one company that's hugely popular in the tiny house world for providing a toilet that has the best of all worlds. It looks like an actual toilet, and it comes with a handle on the side that churns the contents around and accelerates the drying and composting process. There are a number of different incarnations of this style of composting toilet. Some have an external vent mechanism that uses a small, low-voltage fan to continually push air through the composting chamber and out of the house. This venting aids not only the overall ventilation of the house, but also in the drying of the material in the composting chamber.

A different toilet option is an **incinerating toilet** ($1,800 plus cost of liners), which is what I opted for in my first tiny house. Instead of having a composting chamber where the contents are deposited, this toilet has an incineration chamber. To use the toilet, you put a paper liner into a stainless steel toilet bowl. When you're done, you push a paddle with your foot, and gravity drops that paper liner with its contents down into a burn chamber. You push a start button, and the toilet starts a burn cycle. In my 110-volt version of this toilet, that burn cycle runs for a full 90 minutes. As that burn cycle runs, there is a vent that expels the smoke and any odors to the outside of the house.

This Incinolet is my incinerating toilet. It's high tech in the bathroom. No need for cumbersome plumbing.

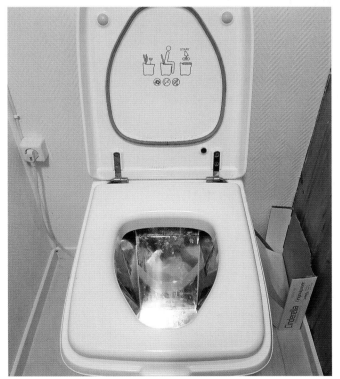
Your tiny house toilet may be more high tech than a normal home toilet—this incinerating toilet certainly is.

The only "plumbing" that an incinerating toilet needs is a way to vent to the outside. It's as simple as cutting a hole through the side of the house.

There are several downsides to an incinerating toilet. They are expensive; I paid over $1,500 for mine. The paper liners are an ongoing cost. And they use a lot of energy to run, making them unsuitable for most solar or battery-powered homes. The benefits of incinerating toilets are that they do not require plumbing, they leave minimal residue in the form of powdery white ash, and they only need to be emptied every week or two.

There is one further toilet option available called a **dry flush toilet** ($600 plus cost of liners). The way that it works might sound familiar to those of you out there who have children and may have used a certain diaper disposal device. In a dry flush toilet, you do your business into a chamber that is a kind of cartridge with a liner in it, almost like a garbage bag material. When you flush, that material closes in on itself and twists, creating what looks like a sausage link. This repeats with each flush until the cartridge runs out. You can flush up to an average of 17 times, and then you need to replace the cartridge. Dry flush toilets are not as expensive as incinerating toilets, but the continual cost is high because of the cartridges

that you need to buy. Energy usage is relatively low in a dry flush toilet, so you could conceivably have one in an off-grid scenario. You will have to dispose of the rather lengthy and somewhat complicated sausage that gets created. This can be done via any standard municipal garbage pickup, since the waste matter is completely enclosed in the liner.

In the end, toilet choice mainly comes down to what you can adjust to, your monetary budget, and your energy budget. If you want the least expensive toilet possible, and you're not squeamish, a composting toilet is the best solution. If you have electricity available where your house is parked, perhaps an incinerating toilet is the better option. If you want the best of both worlds, try a dry flush toilet. And if you have sewage system access, a traditional flush toilet might be the way to go.

A dry flush toilet looks pretty much like the toilets we are used to in traditional houses.

INVITING FOLKS OVER: ENTERTAINING

People don't generally get into tiny living because they loathe entertaining, don't have any friends, and never want to invite anyone over. Quite the opposite: in my experience, people in the tiny house community tend to be very social, super friendly, engaging, and eager to hang out and share stories. There is a real sense of community in the tiny house world that comes from doing something that's out of the mainstream.

So, if you have a tiny house, you're going to want to invite people over. But first you have to get rid of the preconceived notion of entertaining in a traditional house. Picture Thanksgiving or Christmas, when large groups of family members come together to celebrate and socialize. This works great in a traditional house setting. You can grab extra chairs from the garage and move tables together; you've got plenty of space. But tiny houses are much more restricted. I have seen tiny houses that have unique

When you have a tiny house, you'll make new friends. You may wind up with some new friends simply by stopping to get gas with the tiny house. People gravitate to them like a magnet.

ways of creating interior space where you could potentially have five, six, seven, or even eight people sitting at a smallish table. But this requires a very well-planned space and just isn't feasible in the average tiny house. My suggestion for entertaining in a tiny house is to **focus on the exterior** of your house: entertain outside.

This is way better than sitting and relaxing in any living room that I'm aware of. A glass of wine and feet in a cool mountain stream, right outside the front door of the tiny house.

Entertaining at a tiny house is more like a **backyard barbecue**. If you approach it from that perspective, entertaining in and around a tiny house becomes much easier to envision. The focus is on the outdoor space, and cooking for a large number of people will also be better done outside, whether you have a barbecue or a fire pit. Have a luau; put a pig on a spit. Rent or borrow some extra lawn chairs and a gazebo, or pop up a tent for inclement weather. Bring lawn games; have a potluck. Try whatever format you're into.

I highly encourage you to invite folks over, especially if they have never experienced a tiny house before. Be ready to answer a lot of questions and share your story. You'll definitely see some envy in your guests—people will be intrigued by your highly personal, hard-earned space as well as by your newfound perspective. But it's a great opportunity to educate a willing audience about tiny house life. I have no doubt your first tiny house gathering will be a great success.

By the way, don't forget to make sure you have enough resources for all your guests to use the bathroom and wash their hands, etc., during a visit to your home. You don't want a restroom disaster!

Yep, this is where I cook. It's great. Over-the-fire cooking—that's how it was done before we were all spoiled by the convenience of electricity.

If you're going to entertain a large group, you're going to have to do it outside. This is nothing new to the tiny house community—I took this shot at a communal meal at a tiny house event.

MOVING AROUND: TOWING

Whether you intend to tow your tiny house a lot, a little, or barely at all, there is almost certainly going to be a time when your tiny house is going to need to make it from point A to point B. You could hire someone else to tow your house for you, but, as usual, it's best to be as informed as possible about the whole situation. Chances are you'll be looking to haul your house yourself, so let's talk about it.

Before setting off, you will have to get your house **ready for towing**. You'll need to find a way to secure everything that is loose or can open in your house. One thing that I found useful is employing child locks. Most big-box stores sell some form of child safety mechanism, and this helps keep drawers closed. Move anything that's up high down to floor level or otherwise secure it. Place furnishings on rubber mats to keep them from shifting and sliding. Remember, when you're towing a tiny house, you're subjecting the house and the contents of that house to quite a lot of powerful forces.

Towing behind a U-Haul looks like this. Many trucks have the welded-on 2" (5cm) ball. If you need something different, make sure the truck has the ability to add a different hitch like this one (and has a brake controller).

Child locks are a good way to secure doors and drawers before you pull away in your tiny house.

Houses are heavy. Your towing vehicle needs to have the **horsepower** to pull the weight of your tiny house (more on this on page 59). Different towing vehicles have different tow rates that are based on the size of the vehicle, the size of the engine, and the hitch configuration on the back of that vehicle. Any old truck is not necessarily going to be sufficient for towing the weight of your tiny house. Smaller vehicles, such as family sedans, don't have much of a towing capacity, since they're not really built to tow things. Larger trucks, particularly diesel trucks, can haul heavier weights.

The tow vehicle's engine is going to have to work quite hard to move the extra weight around. Make sure that the **mechanical state** of the tow vehicle is good, because any mechanical deficiencies are going to be highlighted during the extra stress created by towing. The power required will also lead to a much higher-than-normal **fuel consumption** for the towing vehicle. Impacts on fuel economy can be severe. I have seen fuel economy cut in half when towing. My truck, which gets around 10 miles per gallon (4.3km per liter) in the city and 14 mpg (5.6km/liter) on the

This Smart car really suits the tiny house. But you certainly aren't going to be using it to tow the house.

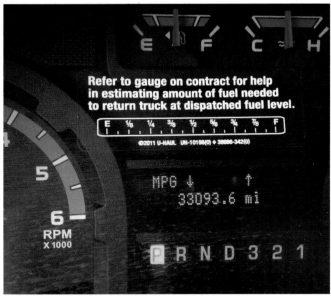

Watch those gauges. Take care of your tow vehicle. You do not want to break down while towing a tiny house.

highway, drops to under 10 mpg (4.3km/liter) on the highway when towing. As with most things, there are a number of factors that impact this. If your house is tall and has a large, broad, flat surface facing the direction of travel, that will create a lot of drag, which hurts fuel economy. You will also see different fuel economy when you are driving across long, flat stretches of road versus driving over mountain passes or similar.

Towing will also cause the tow vehicle to generate extra heat, which will have an impact on the **engine oil** and the **transmission oil**. It is an excellent idea, if your tow vehicle doesn't have it, to have a **transmission cooler** installed (may cost around $250). This is an extra radiator that the transmission fluid gets pumped through, creating a constant and quicker cooling of the transmission fluid. If you can keep the transmission fluid cooler, it helps prevent damage to the internal parts of the transmission. As for the engine oil, if you are used to changing your oil every 3,000 or 5,000 miles (4,800 or 8,000km), you're going to want to **change your oil** more often if you're doing a fair amount of towing. Stay on top of these factors.

Towing a tiny house forward is one thing; stopping the tiny house is another. **Electric brakes** and a **brake controller** are essential. Most tiny house trailers, unless they're for micro tiny houses, have some form of brakes on them. The most common variety that you'll see is an electric brake trailer. Electric brake trailers connect via a wire and a seven-pin connector into the back of the tow vehicle. That electrical connection controls not only the brake lights and various lighting options on the

This kind of situation creates a hard braking scenario. Trailer brakes are key here. Without them, the weight of the trailer will push your tow vehicle right into an obstacle.

trailer, but also the brakes themselves. You need a brake controller in the tow vehicle to be able to apply the brakes on the trailer at the same time that you are applying them on the tow vehicle. With electric brake trailers, instead of having a considerable weight behind you that pushes onto the tow vehicle when you hit the tow vehicle's brakes, the trailer itself brakes, too. In other words, there's no additional burden on the tow vehicle. Simultaneous, separate braking like this is essential on larger trailers, since your tow vehicle not only has to stop itself, but also a 10,000- or 15,000-pound (4,500- or 6,800-kg) tiny house behind it. The electric brakes make safe and quick stopping possible. There are some weight scenarios where you can get away without trailer brakes, though. As an example, the truck I have can tow upwards of 10,000 pounds (4,500kg) with trailer brakes. My truck can therefore tow a smaller structure of, say, 2,000 pounds (900kg) without trailer brakes; the truck itself weighs

about 7,000 pounds (3,100kg). That same tow weight of 2,000 pounds (900kg) behind a much lighter vehicle, however, would not be suitable to be towed without trailer brakes.

It's important to understand before setting out that towing is not just like driving. Go slowly until you are comfortable with how it feels to tow a large, heavy thing behind you. Understand that you're not going to be able to drive at the speeds you're used to going down the highway. In order to be safe, you will need to go slower, because the faster you go with the heavy weight behind your tow vehicle, the more potential for danger you're introducing. Remember that you have a much longer stopping distance when you're towing thousands of pounds. Think about safety; think about the other people that you're sharing the road with; and keep these things front and center in your mind when you're towing a tiny house.

MICHAEL FUEHRER

Michael: My name is Michael Fuehrer, and I live in a bus named Navi. I got into bus life and tiny living after graduate school. I essentially had the desire to travel, but also not just live in a car. I didn't want to have to put down the back seat of a car and things like that. I wanted to be able to have people over, have a space that I could feel at home in, that I could live in. I started searching online for options for tiny homes. That's when I thought a school bus felt like it was good. A lot of people ask me why didn't I buy a motorhome or something, because it's similar. What I say is that I wanted it to be my home and I wanted it to be my design. So I ended up going with the DIY school bus and built the whole thing out.

Michael's Navi Navigation to Nowhere bus tiny house conversion.

Chris: *From the first moment where you got this concept in your head, what has been the thing that's been most surprising to you?*

Michael: I think the community. Little did I know that there's this large tiny house minimalist community that I'm now involved in. So it's kind of fun that it ended up opening up into this larger world I didn't even know about when I was building my bus. When I first bought my bus, I didn't think I was the first one to do it, but I didn't know anyone else who had.

Chris: *What's one piece of advice that you have for people looking to get into tiny living?*

Michael: Never give up. I feel like I encourage people to just go for it. There's a lot of times you hear the dreamer mentality of, "Man, I wish I could do that," or, "I really want to do this," or, "I can never do that." It's one of those things where I had the same thinking when I was in school: "I really want to do that, but I need to finish school first." I've never looked back since I did it. I could very well be doing a nine-to-five job, and I would probably be fine. But I think there's another level of happiness I found, and I try to encourage people to start taking steps within their lives to start doing it, rather than just continually thinking about it. Obviously, research and time are important in your process, but start doing the research, start doing the process, and actually take steps.

Chris: *So taking the first step is important.*

Michael: Some people want to do a travel lifestyle, and if that's the case, I always suggest people start looking at their work or what they currently do and maybe try to go remote before you start. Because if you can do that, then you're going to take that factor out of the equation. If you're not trying to travel, and you're trying to be more stationary, start making life decisions with your finances. Maybe right now you have two car payments and really high insurance or two rentals. Whatever it is, just start looking at your actual budget and seriously taking a look at where you're spending your money.

Chris: *Fast-forward a year. What do you see yourself doing?*

Michael: It's going to be something to do with alternative living styles, even if it's homesteading or something else. I don't see myself going back to the way I lived before, but I'm not exactly sure where this will take me. I think right now I'm in a very free place. I have this bus that I own outright. I'm a single guy. I have a lot of freedom of choice. And I'll always be ready for the zombie apocalypse.

The inside of the bus.

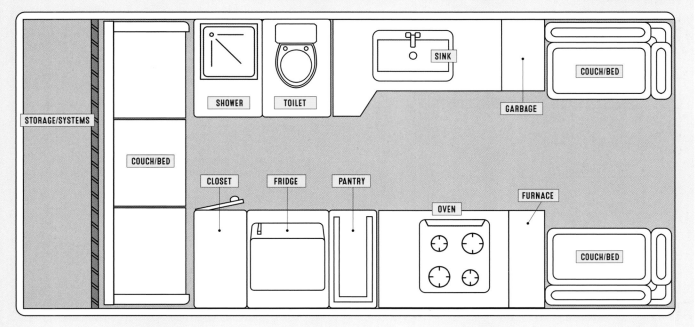

Michael Fuehrer's bus floor plan.

THE TINY TRAVELING THEATRE

Jesse: My name is Jesse Dufault. I'm one half of the Tiny Traveling Theatre. I'm an actor, musician, cinematographer, and photographer. Our dream was to create this tiny house pop-up art venue so that we could bring storytelling, film, music, theatre, and education to places that wouldn't typically have access to it.

Nora: I'm Nora Eschenheimer. After college, I went to Europe as a solo female traveler and started living out of a backpack. After all this traveling, I realized that my two passions in life were being on the road and acting. I wanted to combine those two things, and that's when I started to come up with crazy ideas for a tiny traveling theatre.

Jesse Dufault and Nora Eschenheimer are the full-time residents and owners of the Tiny Traveling Theatre. Visit them at *www.tinytravelingtheatre.org*, or on Facebook and Instagram, @tinytravelingtheatre.

The Tiny Traveling Theatre with its built-in stage.

The theatre in action. Pull up a chair and be entertained. Nora and Jesse have a unique structure that brings entertainment wherever it needs to go.

Chris: *How did you end up creating this amazing structure?*

Jesse: The first iteration of the Tiny Traveling Theatre was on the back of a Denny's napkin. We then drew up some serious concept designs on graph paper, and it took over a year of designing and planning until we felt confident.

Nora: Fortunately, Jesse's father is a retired master house builder. And my father works in heavy machinery, so when it came to anything that was steel, my dad was like, "Don't worry, I got this."

Jesse: What I loved about my dad's help is that he always embraced what we wanted to do. He never said no to us. I think that's what we thank both of our fathers so much for, for saying, "I know what you want; we're gonna make it happen."

Nora: You have to seek out those people in your life. It's all about support.

Chris: *Is there anything that you would do differently the second time around?*

Nora: We had this trailer custom fabricated. We were told that welding together was going to cost $2,500. We thought that was a bargain. The part that we didn't wrap our heads around is that that didn't include the axles, the wheels, or the everything else. It would have been a lot more cost-effective, faster, and less stressful to buy a complete and fully outfitted tiny home trailer.

Jesse: We also have a ladder system to get to the loft, and I'm not sure I would go that route again. Most people in 20' (6m) tiny houses have the bookshelf that goes up and has stairs, and then your dog can run up. But we went with a ladder, and when I wake up, and I have to go to the bathroom, I wish there were a slightly easier way to get down from the loft. It's not the end of the world, though, if that's the worst thing we've discovered in the past seven months of living in the house full time.

LEGALITIES AND MORE

Although this section is at the end of the book, knowing where you are going to park your tiny house is good to figure out long before your house is built. There is a lot going on in various parts of the country that is changing the legal landscape around tiny houses. National groups like the American Tiny House Association are educating and advocating for easing restrictions for living in tiny houses. In many parts of the country, this is still an uphill battle. Nevertheless, the prognosis is promising.

EVERY HOUSE NEEDS A HOME: WHERE TO PARK

Legalities around tiny houses are a tricky subject. For one thing, legal restrictions and requirements are very local. Your particular situation is going to depend on where you live, but you also have to consider that you may be moving your tiny house across county and state lines. Unfortunately, there's currently little commonality across the United States when it comes to rules and regulations that impact tiny houses. However, there are some things that will generally hold true regardless of where you are in the country. Read this section carefully, but take none of it as gospel—do your own research based on where you want to place or take your tiny house. There is a relatively new site that is working to catalog all the various places to park and live in a tiny house: *www.searchtinyhousevillages.com*. More resources like this are starting to be developed online.

Finding Space Legally

Tiny houses are generally considered **custom recreational vehicles** in the eyes of the law, so whatever laws pertain to recreational vehicles, broadly speaking, pertain to tiny houses as well. This isn't true everywhere, but it's generally true. If you've gotten this far in the book, you've hopefully already given some thought as to where you're going to put your house.

Even in a remote area where you think you won't be in anyone's way, you have to contend with the local regulations—you can't just park and stay anywhere.

So why is it so difficult to live legally in a tiny house? Many municipalities are somewhat wary of tiny living for a number of reasons. Since most tiny houses are built on wheels, they are afraid of the **transient nature** of tiny living. They don't want people moving in and out of their town and creating a type of tent city that's going to be frowned upon by the permanent residents of the area. **Property values** come into play; people, rightly or wrongly, worry that the presence of tiny houses will negatively impact their property values. Another critical aspect is **revenue**. How are towns going to make revenue from houses that are on wheels? Towns usually make revenue from property taxes. Those taxes then feed into the local economy and pay for schools, police, fire departments, and other municipal services. How do you properly tax a tiny house? It may not even be possible. All these considerations are difficult for municipalities and towns to deal with, so more often than not they hit the easy button and say no to tiny living in general.

That said, some towns are starting to embrace the existence of tiny living. In the United States, the movement seems to be strong out west and slowly heading east. One town at a time, one experience at a time, townships are opening up to tiny homes. And there are actually some advantages that tiny house presence can afford to towns. One major advantage is that tiny house living can help a town address an affordable housing crisis, which is common across the country. Some cynical people might say that towns don't care whether or not they make affordable housing available, and this may be true in some places. In general, though, having a solution for affordable housing is good PR for a town, and a tiny house solution is one way of addressing that.

Legally, you have four general options when it comes to parking your tiny house.

YOUR OWN PROPERTY: Obviously, you can almost certainly park your tiny house on land you own. This isn't a given, though, so make sure you check the local laws. Some townships allow accessory dwelling units (ADUs). An ADU is an additional structure that you're allowed to have on your property. ADU regulations were not necessarily developed for tiny houses, but they can assist them dramatically.

If you own property, you can likely build or park permanently in your own backyard—but check your local zoning laws first to be sure.

ZONED LAND: You may be lucky enough to find a town that is open to tiny houses, at least in certain areas.

MOBILE HOME PARKS: Some mobile home parks, also known as trailer parks, are making space available for tiny houses. These parks were created before the tiny house boom and were designed to accommodate pre-manufactured homes. Converting these rather drab and dreary cookie-cutter places into something nicer has many benefits. The mobile home park can charge a bit more for the additional landscaping and the community areas that groups of

A mobile home park might be a perfect fit for your tiny house, on a temporary or permanent basis.

tiny houses typically create among themselves. These are all positives for mobile home parks, which is something you can point out to a park you're trying to convince to let you park in. Prices will vary widely based on location and amenities, but figure a few hundred dollars per month in "rent."

CAMPGROUNDS: If you live in a rural area, there are generally going to be RV campgrounds around. Some campgrounds are making tiny living an option, allowing people to long-term lease space. This creates a year-round revenue stream for campgrounds that might otherwise see serious dips in patronage over the winter months. These campgrounds are already equipped to handle power needs, sewage needs, and often laundry needs, making them a great fit for tiny homes. Again, prices will vary based on location and amenities, but a few hundred dollars per month is to be expected.

Camping and tiny living aren't that far removed from one another.

Finding Space in the Gray Area

Another idea for finding a tiny house living space, which has worked for many people, is to create a **barter situation**. When you find someone that needs help, whether it's with house-sitting, farming, construction, or something else, you can barter some of your time and energy in exchange for a spot to park your tiny house. Of course, this requires you to constantly search for a new barter situation when your current one has run its course. This can be tiring, but it can also be exciting; it really depends on your personality and what you want out of tiny home living.

In the absence of designated spaces for tiny house living, some people simply go forward with tiny living regardless of the legalities. Doing so is up to the

If you know someone with space to spare, perhaps you can barter a service to be able to spend some time parked on their property. This is what Alex Eaves has done (see page 44).

individual. It is not possible for me to endorse doing so or say that it's a good idea, but many people are throwing caution to the wind and putting their tiny houses in places where they don't necessarily belong. Often this sort of thing only becomes an issue when someone complains. If you can find a place to put your house where it's not **visible** to anyone else, not **encroaching** on anyone else's space, and where you're not likely to **impact** anyone else's property use, you may never have a problem. It is unlikely that people will complain about something that they can't see and don't know about.

If you are in a situation where finding a legal place to park your tiny home is difficult, there are some sources I would encourage you to seek out. There are **tiny house groups** out there. I head up a chapter of a *Meetup.com* tiny house enthusiast group in New York City and Philadelphia. There are others in different parts of the country. These are also excellent resources even if you don't need a place to park, because you will have support on any tiny house subject you might desire. These kinds of groups are a great gathering point for collective voices to help facilitate changes within the local area. I happen to live on the East Coast, in the very dense suburban area outside of New York City, so the prospects of full-time tiny living near me are minimal (at least currently). But I'm working to change that, and I encourage everyone to do the same.

Tiny House Expedition (*www.tinyhouseexpedition.com*) is another great source for keeping up to date on the legalities of

These van decals aren't meant to camouflage this tiny house on wheels, but the more you can blend in and not be seen, the better.

tiny living. The project is run by a tiny-dwelling documentary filmmaker couple. They also have a YouTube channel where you will find their impressive and helpful videos, including a documentary called *Living Tiny Legally*. They document and highlight the success stories around people going tiny and the municipalities that are coming on board with the concept and allowing tiny living in their jurisdictions.

Basically, get creative! Focus very local and find out what specifically applies to you and your situation. Be diligent about it; be vocal about it. Find like-minded people who are willing to help you. Remember to be flexible and not too choosy. The information you want is generally not going to be readily accessible (yet); it's going to involve some legwork on your part. I wholeheartedly encourage you to seek it out. And when you do find it, share it with others who are facing the same situation.

The bottom line is, this book cannot possibly address your particular situation. I encourage you to look at the legalities of living in a small space in your area. And remember to look into the regulations in surrounding areas. What is true in your particular town or location may not be true five or ten miles down the road.

PROTECTING YOUR INVESTMENT

The tiny house boom is a relatively new phenomenon, and this has left the insurance industry trailing behind a bit. Insurance has always been available for standard recreational vehicles, but because tiny houses fall into more of a custom recreational vehicle category, it can be somewhat tricky to insure them. The main reason is that tiny houses are built by a lot of smaller, independent entities, whereas standard RVs are mass produced by large companies and have model numbers, serial numbers, and all other sorts of identifying features. The problem with this for the insurance industry is that it is hard for them to place a value on unique and highly customized tiny homes. It's also hard for them to have accurate insight into the standards to which they were constructed. All of this makes tiny houses much harder to insure.

One way to make it easier to insure your tiny house is to have it built by a larger, more established tiny home manufacturer, because they have specific standards that they adhere to. As discussed in a previous section (page 84), you can get an **RVIA certification** for these kinds of tiny houses. Of course, the tradeoff here is that buying an RVIA-certified tiny house will most likely cost you more,

RVs and tiny houses built by reputable manufacturers may be less customizable, but they are more easily insurable.

and you will also have less ability to customize it to your specific needs and wants.

Despite growing pains, the field of insuring tiny houses is expanding, and it is usually possible to get tiny house insurance. Just be prepared: the insurance may cost you more if it's difficult to determine the nature of your house, how it was built, and who built it. Insurance companies inherently seek to avoid excess risk and will, by default, charge you more for a tiny house. They will ask you how often you travel with your house and where. Many will offer a plan that allows you to purchase insurance only for the times that your house is on the move. One offer I remember getting was $250 for each trip in the house. If you average it out on a per-square-foot basis, insuring a tiny house is a lot more expensive than insuring a regular structure. But as competition increases, I expect that the cost of insuring a tiny house will come down over time.

No one is forcing you to insure your tiny house, though; you could go without insurance. There's a risk involved with that, but you can still cover elements like your liability through your car insurance or an umbrella policy.

Protection Beyond Insurance

So what else can you do to help protect your tiny house, whether or not you have managed to insure it? Mainly you will want to prevent theft of the tiny house. This will depend somewhat on where you're parking your tiny house. Is it a remote location? Can it be seen from the road? Is it likely to attract attention? These are elements that could impact how at risk your house is.

There are several means that you can use to help keep your tiny house safe. **Hitch locks** ($50) are a popular device; they prevent someone from easily hooking your house to the back of their truck and driving it away. You can also employ things like

This is a hitch lock. It's not a guarantee to prevent a tiny house theft, but it will take few extra minutes for a would-be thief to circumvent it, so it's a pretty good deterrent.

remote cameras ($300) that are motion activated. They send you an immediate notification when someone is doing something to your house or near your house. Another option that's become viable over the last few years is **GPS tracking** ($150 setup plus a monthly fee). You can hide a sensor somewhere in the house where it would be difficult to find. Then, if the house is stolen, you can track it.

Keep in mind that would-be criminals likely know about all these protective methods. However, it's been my experience that there is minimal theft of tiny houses taking place. A tiny house is very conspicuous when you're towing it down the road, and many tiny houses are truly unique with easily identifiable features. For these reasons, stealing a tiny house is not something that many criminals are keen to do. Smaller things like cars are easier to steal because they can be hidden in shipping containers, which also block GPS signals. Typically, tiny houses don't fit into shipping containers. I used to be somewhat concerned about my tiny house getting stolen, but over the years I've grown less worried; I just don't see it as something that's likely to happen.

PART-TIME TINY HOMES

Not everybody necessarily wants a tiny house to live in full time. Tiny houses tend to also serve very well as part-time secondary homes. I want to dedicate some pages to this aspect of tiny living. It's good to view a tiny house as something that's much more versatile than just a means to downsize your entire life. To begin with, these structures are great alternatives to campers, tents, and other ways of spending time away from home. Some people are not into camping in the traditional sense; the prospect of sleeping in a tent is not appealing to them, and they need a little bit more creature comfort to be able to spend time in nature. For these people, a part-time tiny house might be the perfect solution. I like to refer to these types of dwellings as **urban escape pods**.

When you live in the city or in an urban environment, it's often easy to find yourself not spending time outside of the city. Sure, you go on vacations, but the day-in and day-out grind of being in the city can wear on people. Having a place to go can be a very enticing option. And achieving this typically has a much lower barrier to entry than most people imagine. A house that you only spend the weekend in doesn't need to be as big or as full-featured as a tiny house that you intend to live in full time.

Part-time tiny homes have two main benefits: they are **less costly** and **easier to build** than full-time tiny homes. When most people think of having a second house or a vacation home, the inclination is to very quickly conclude that you don't have the money for it. But we're not talking about a $250,000 condominium in Florida—we're talking about a much smaller structure that costs in the $8,000 to $18,000 range (half that if you are going to build it yourself). Regarding the construction, while you might be intimidated by framing and finishing

There's nothing urban, crowded, or noisy about this this part-time getaway.

out a 13' 6" (4.1m) tall house on a 26' (8m) trailer, it's much easier to see yourself building a smaller, lighter 6' x 10' (1.8m x 3m) house on wheels. Trailers for such structures are abundant and readily available. My company produces such structures on standard utility trailers that are both sturdy and reasonably priced (around $8,000 to $15,000). The build process for one of these can be completed in a few weeks with materials available in most larger hardware stores. A hands-on instructional build of one of these is the subject of a book I'm working on at the moment (look for it in late 2019).

There are several areas in which part-time tiny houses are similar to full-time tiny residences, but with key differences. Let's explore them here.

Location: You will need to find a piece of land where your part-time tiny home can reside. It doesn't have to be land that you necessarily purchase. It could be land that you barter for or rent from somebody else. This land would ideally be far enough away from an urban environment to not be excessively expensive, but close enough to be able to get to rather quickly. Zoning will be a bit more relaxed regarding placing your structure as well,

A part-time tiny house doesn't have to be fully equipped or expensive. It can even be a good way to put some of those extra decorations and kitchen utensils you've stored away to good use!

since the structure is not a full-time residence. Either way, be sure to check into the local restrictions around spending time in and parking a recreational vehicle, as that's likely what your part-time tiny house will be considered.

Water: Water is not necessarily a big problem in a part-time tiny house, because you can bring water with you, and if you're only spending a day or two at a time in the tiny house, camper, urban escape pod, or whatever you want to call it, then you will not have a need for large amounts of water the way you would in a full-time residence.

Waste: Where and how are you going to go to the toilet? This is also not a huge deterrent for part-time tiny homes. The easiest thing that you can do is get an RV toilet and use that for the brief amount of time that you're spending out in the structure. An RV toilet is a toilet that uses a water tank and a holding tank. You use a small pump mechanism to flush the toilet. Fancier toilets of this kind have an electric pump, but that makes the device more complicated and prone to breaking, in my view. Stick to the simple hand pump if you decide to go this route. This toilet can be used numerous times before it needs to be emptied or refilled with water, so it's quite convenient for short periods.

Find a location that will give you the kind of escape you desire.

Speaking of water, imagine being able to escape to a watery heaven like this!

The more remote your location, the better the escape, but the harder access will be without a car. Figure out the sweet spot that's right for you.

There are also some areas that tend to be less of a consideration for full-time tiny residences that are more important to consider for part-time tiny residences.

Access: You will need to be able to travel back and forth between wherever you live and wherever your urban escape pod resides. If it's an ordeal to get to, you won't want to go, and that defeats the purpose. You don't necessarily want to be towing the structure around all the time, either. If you are going to haul the structure, you need to realize that you have to have a suitable vehicle. This may or may not be feasible in your city, and if you are merely going to be traveling out there to spend time, it may make sense to just rent a car or use some kind of car sharing service. You may also be lucky enough to find a piece of land that is somewhere near a train or a bus route so that you can utilize public transportation to get to (or at least close to) where your tiny dwelling resides.

Renting: Once you have a part-time tiny house, realize that if you're not in it all the time, you can potentially rent it out. Renting can be done through a variety of different websites, such as Airbnb. There are also sites such as *outdoorsy.com*, which are specific to renting out campers and smaller structures. You can also allow friends and family to use your structure. Subletting your space is not necessarily lucrative, but it can cover and defray your costs rather effectively. Just make sure you aren't breaking any local laws and are declaring what you need to on your taxes!

Value: Keep in mind that, contrary to buying a vacation condominium in a lucrative market, a part-time tiny house (like any tiny house) will generally depreciate in value over time. A part-time tiny house shouldn't be viewed as an investment vehicle in the traditional real estate sense. But so what? It didn't cost that much to begin with. People buy cars all the time that become worth less and less as time goes by. If you get enjoyment and relaxation out of your part-time tiny house, then that's what makes it worth it.

In the end, being in nature is just simply nice. If you want to have an upgraded experience via something like an urban escape pod, then a part-time tiny house is perfect for you. You really can have relative luxury in a rather compact package.

Beautiful shots like these will make people want to come stay in your tiny home, if you're interested in using it to make some money on the side.

Your part-time tiny home is for you to escape to now, not for you to invest in and sell at a profit later.

Is there anything more peaceful than a retreat like this?

THE JOY OF TINY HOUSE LIVING

OTHER TINY DWELLING USES

I view tiny houses, and tiny structures, as "extra space" in its most basic form. Because of their extreme customizability, they are essentially blank slates built on a set of wheels. The possibilities are pretty endless. In this section I am going to highlight all the purposes for which people have asked me to build them a space on wheels. The idea is to broaden your perspective regarding what a tiny house should be versus what a tiny house can be.

All of the various uses that I'm about to list here are for very small, highly mobile spaces on wheels. Any number of these options can be made larger as well, but for the most part, the uses make the most sense in smaller, perhaps 50- or 60-square-foot (4.5- or 5.5-square-meter), structures.

Backyard office: A tiny house as a backyard office is a great idea. When we live in traditional spaces, they don't always allow for integration of office space. It's also so much easier to get distracted by all the various things around you. A backyard office could be as simple as putting a desk and a chair and whatever else you need inside of a tiny structure. Perhaps clients or customers could even visit you in that space, in which case you could have a little bit of a seating area for them. A backyard office can look any way you want, and if it keeps you away from those distractions that are part of your actual house, then it's truly serving a purpose.

Guesthouse: Most of us like having guests, but sometimes we live in spaces that aren't conducive for having people visit. Having a guesthouse tiny house in the backyard is an excellent way to go. The guesthouse should feature a place for somebody to sleep, with some privacy, but it doesn't necessarily have to have a full kitchen or bathroom. Those could be shared resources within the actual house. It

All that fits in this trailer tiny house is a bed and a (rather large) TV. It's perfect for giving guests a comfy place to lay their heads and an extra bit of privacy.

could just be as simple as having a comfortable and private bed.

Pool house: If you're fortunate enough to have a pool in your backyard, why not create a tiny house as a pool house? This could be used for changing, for storing towels, for keeping a fridge and refreshments, or all three. You don't need a large space for that sort of thing, and having the dedicated extra space right by the pool can prevent people from needing to walk into your actual house with wet feet and wet bathing suits.

Media room: Perhaps your kids like playing video games, and there's competition for the television. You could dedicate an extra space in the backyard for either you to watch movies or for the kids to play their video games—as loudly as they want.

Exercise space: In particular, a yoga space would be a very simple tiny structure. You mostly need four walls, a roof, and a yoga mat on the floor. The privacy while you're huffing and puffing (or meditating) will surely be welcome.

Crafting space: Many people love crafting, and breaking out your crafting supplies and putting them away again can be cumbersome if you craft frequently. It's great to have a dedicated space for activities like painting and scrapbooking where you won't annoy anyone by taking over the kitchen table for hours (or days).

Treehouse on wheels: Small structures can make a great play space for kids. You can turn a tiny house into a sort of a pirate ship, or the kids can use it for sleepovers.

Recording studio: If you're a musician, perhaps you could benefit from an outdoor recording studio or a place to practice and play your music without bothering the rest of the family.

Tailgating space: Sports fans who go to events know how popular tailgating can be. What if you could bring a small, dedicated tailgating space with you instead of just popping open the trunk of your car? Some people have done this with school buses and vans. You can even have a structure with

Yes, a tiny house can be designed for binge-watching Netflix. Just look at the entertainment system in this tiny house!

This tiny structure fits perfectly into a parking spot as if it was made for tailgating.

an integrated TV to watch the game from the parking lot. You can use it to store the various things that you need for tailgating, such as coolers and grills. A tailgating structure can be fairly basic and doesn't need to be fully insulated.

Vending space: Sell your crafts or artwork from a completely custom vending space that allows your work to truly shine. You'll have to check the rules depending on the shows you attend, but this can be an attention-grabbing way to bring goods to your customers and customers to your goods.

Food truck: Sell food or other consumable goods out of a dedicated mobile tiny house—with the proper permits, of course.

Looking to travel to art fairs and sell some kind of craft? Tiny houses make great vending spaces on wheels.

A house as an ice cream truck—or is it an ice cream truck posing as a house?

THE JOY OF TINY HOUSE LIVING

Church: Why not take a church or other place of worship with you? It gives new meaning to the phrase "destination wedding."

Event space: If you have a company that holds lots of events, or just want to be entrepreneurial, create a traveling event space for any of a variety of functions.

I'm constantly amazed at what people come up with. I've seen a pet sitting business create a space where different pets from the same family can spend time together when they are being taken care of away from home. I've seen marketing companies turning to tiny houses to promote clients' goods at various events. I've seen tiny houses used for disaster relief and transitional housing for the homeless. The possibilities are pretty much endless. What other uses can you think of?

A tiny chapel on wheels.

This is an event space for corporate functions and other purposes that could benefit from having a dedicated space on wheels.

NEXT STEPS: MOMENTUM

Congratulations: If you're reading this, you've read almost the entire book (probably). After you've worked your way through every preceding section, and ended up here, what do you do next? This is a point where it's easy to lose your momentum and not take your tiny house dream toward a tiny house reality—whatever that reality happens to look like for you. Reading this book has been a great start, but what do you do now? I'll break this down into three broad topics that you can tackle next, and then cover them one at a time in detail.

- **Research:** Increasing your knowledge via online resources or other books (like the one that you're holding in your hands right now).

- **Planning:** Drawing out your tiny house plans in technical detail.

- **Exposure:** Seeing and touching as many tiny houses as you possibly can.

Research

Books are great. I thank you for buying (or borrowing) mine and for giving it a look. **Books** are a super valuable source of fact-checked, reliable information on a whole host of topics. Now, that said, there are a ton of online resources that you can tap into as well, as I'm sure you're aware, and that's what I want to focus on here.

I find that the inspiration I get from what other people do often comes from an online forum. The places that I like to go most are **Instagram** and **Pinterest**. Those are the two most visual platforms, **Facebook** being a close third in that regard. What these platforms allow you to do is to get a sense of what appeals to you and what doesn't appeal to you rather quickly. On a platform like Pinterest, for example, I suggest that you start putting together a couple of different boards that are a central repository of the various things that you see and like out there on the web.

There are also a lot of tiny house tours on **YouTube**. I've been fortunate enough to have had some of my tiny houses featured on prominent channels. These tours are a great and realistic way to get to see a little bit more detail than you do from just pictures on the sites that I have mentioned.

These various sites will help you to ascertain what your style is and what appeals to you. Without having a good sense of this, it's tough to move on to the next step in the process.

I also encourage you to read **blogs**. Many tiny house builders and tiny house dwellers have blogs. These are typically more focused on the day-in and day-out reality of tiny houses. They will give you a wealth of things to think about in terms of what your tiny life might look like.

Last, but not least, don't forget about **Google**. When faced with a specific question about solar panels or when you're starting to think about what

A YouTube tiny house tour is much like a physical tour... but without the lines!

tiny house trailers look like, those are good things to put straight into the Google search engine. It's a highly useful tool and will return relevant links for you to dive deeper into what you're looking for. The free availability of information on pretty much any topic has made a lot of the movement of homegrown tiny houses possible. How else would it be that so many people get to the point of feeling empowered to tackle something like building a house?

Next Steps: Momentum

Planning

After you've exhausted all the various online resources and feel like you can't cram any more information or inspiration into your head, the next step is for you to come up with more of a plan. I broadly call this "**drawing it out**." What does this entail? Drawing it out is really about putting your ideas together and creating something tangible. When I say draw it out, I mean that quite literally—this could be either in a two-dimensional fashion, which could take the form of a sketch, or even a three-dimensional fashion, if that's how your brain works better. Don't worry, you don't have to be an artist. It doesn't have to be insanely beautiful. It's just a great way for you to capture what it is that you want. Using a piece of paper and a pencil is the simplest route.

The next level up from pencil and paper is a program called **Sketchup**. It's not that difficult to learn how to use. You can utilize the program to do more of an architectural rendering of your space. This can be a two-dimensional floor plan, but Sketchup excels if you're looking to create a three-dimensional layout. Also, if you're not particularly good at Sketchup or if you have no desire to learn how to use the software, there are folks out there who can help you, and there are resources that will allow you to turn your hand-drawn sketches into something that is more like a Sketchup rendering. Look for knowledgeable folks on freelance websites; there are any number of people who will do this sort of architectural rendering for you.

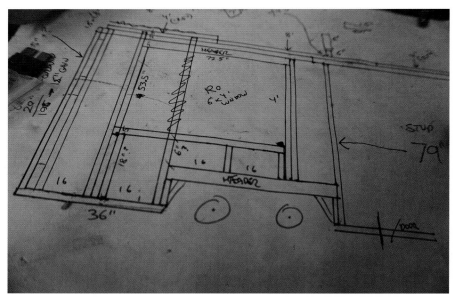

Your first draft doesn't have to be quite as detailed as this—the important thing is to get started.

If making a three-dimensional Lego model of a tiny house helps you plan, then go for it.

While you're doing all of this, try to keep in mind some of the things that you have taken away from this book—things like making sure you aren't putting a door right on top of the wheel wells on your trailer. Think about what kind of roof appeals to you versus what kind of roof makes sense for your needs. Think about color palettes. Think about where and how many windows you want in the structure. This is all a part of drawing it out, to begin to **visualize the structure**. As you lay it out, keep in mind the space considerations. Try to draw everything to scale, and try to have an excellent idea of what will go where. This will keep you on track regarding dedicating space only to the things that you truly want or need.

Exposure

Once you've done some planning and sketching, the third stage is seeking to get more hands-on time with real tiny houses. There are several ways that you can do this.

First of all, sites like HomeAway and Airbnb offer you the opportunity to actually **rent and stay in a tiny house**. I highly encourage doing this, because there is no equivalent to learning about and getting a feel for a tiny house like actually living in one.

You can also look for **tiny house shows**. There are various shows of different sizes that take place all the time across the continent (check out *www.unitedtinyhouse.com*). Even if it's a small show that

You need to get out there and touch some tiny houses—literally. There's nothing like it.

Attending tiny house events will give you the opportunity to see and compare many different types and models of houses.

happens to be near where you live, by all means, go to it. Whether there are three tiny houses or thirty tiny houses, you're going to get a lot of information and a much better feel for what this kind of limited space actually feels like. Attending a show has the added benefit of allowing you to ask builders questions about tiny structures. Shows are an excellent opportunity for you to take pictures and make note of things you like and things you don't like. It will also help you narrow down sizing considerations. You may think that you want a 24' (7.3m) tiny house, but when you set foot in a 24' (7.3m) tiny house for the first time, you may go, "Wow, this is just insanely large, and I'm looking for something smaller than this." This is something that you're only going to find out by actually seeing how someone else has interpreted a small space.

Another route that you can pursue is to seek out **local builders**. I have people reach out to me as a builder and ask if they can come by the workshop to see whatever I am working on. Within reason, I always say yes to requests like that. As a builder, I'm interested in having people be more engaged and having them see what takes place. A person coming in and seeing a house in the build stage will help them visualize the things that they want. It's also another excellent opportunity for those people to ask questions and to get more deeply involved in the process.

The final suggestion I have for you is to seek out a tiny house **building workshop**. What is a tiny house building workshop, precisely? They'll vary based on

whoever is providing the seminar. But they typically consist of a weekend where some folks gather, share a good amount of information, and participate in some hands-on work. Tiny house building workshops of various kinds take place in different geographic locations. This is a good thing to seek out on Facebook or, more specifically, Facebook groups. Search for tiny house workshops online and you'll perhaps see workshops that have already taken place, but that's an excellent opportunity, too: you can reach out to the people who organized those workshops in the past and ask them if they have any upcoming ones (check out *http://relaxshacks.blogspot.com*).

Don't Stop Here

I hope that these different concrete tasks will be what you need to keep the ball moving. I encourage you to stay on this path, and not simply put this book down and be proud of yourself for having read it. Don't lose that momentum!

You want to be where I'm standing: in the doorway of a half-built house, happy in the knowledge that you're actively pursuing your dream.

THE JOY OF TINY HOUSE LIVING

Jenn: My name is Jenn Baxter, and I'm a blogger, an author, and a speaker, based in Charlotte, North Carolina. Most people may recognize me from being on HGTV's *Tiny House, Big Living*. I was on season one.

Chris: *So what's your story? How did you end up where you are?*

Jenn: The way I fell into this whole tiny house world was kind of by accident. I was watching Netflix while I was housesitting for somebody back in 2014, and I happened upon this documentary that was called *Tiny*, about a couple who built their own tiny house on wheels. Up to that point, I had never even heard of tiny houses. I watched this documentary, and pretty much by the end of that 90 minutes I was sold; I felt like this was exactly what I needed to do. I had never really been somebody that was into designer labels, and I didn't like having a lot of stuff, and I had come to a point in my life, after being through several traumatic events, where I was ready for a reboot and a restart. My mother, who was my very best friend in the whole world, had recently passed away at the end of 2013. She had left me a modest amount of money, and I had been planning before this to use that money as a down payment on a townhouse or a condo. When I realized that I could use that same amount of money and purchase a tiny house outright and actually be rent-free and mortgage free, that was just the icing on the cake. I thought, this is a perfect way to also give a little nod to her and honor her. That's how I got into the whole world of tiny houses. I had a 160-square-foot (15-square-meter) tiny house on wheels built by a builder in Raleigh, and I moved into that in 2015. I didn't, unfortunately, get to stay in it; it had some structural issues, so I only lived there for about six months. I'm still very much into tiny houses, still very much support the lifestyle, and now I make it my mission to travel around the country and also teach online and help people to figure out how to downsize their lives—how to get rid of the clutter and the excess so that they can focus on what's most

Photo credit: Jenn Baxter, photo by Jessica Milligan

important to them and live a life that's more joyful and more peaceful and just not so full of stuff.

Chris: *I watched the same documentary; it is pretty compelling, I have to agree. As it relates to clutter, what's your thinking behind why we have so much clutter collectively in the first place? How do we end up here, with all this stuff we're emotionally attached to?*

Jenn: I think over time that stuff just kind of happens, though so subtly that we don't realize it. One day we blink and it's like, wow, I have a garage full of stuff, and I can't park my car there anymore. It's about getting out of that spot where you're blind to your own stuff, and about looking at it with objective eyes and pulling yourself back and saying, "Okay well, why do I even have this in the first place?" It's definitely a mindset shift that you have to make.

Chris: *What role, do you feel, does consumerism and how we're always being marketed to play in this process?*

Jenn: I often refer to the period of my life where I had that switch where I decided that I wanted to start living a minimalist lifestyle as the blinders coming off. We are constantly being told to get more stuff and to have more stuff and to have the newest thing. We think that's what's going to make us happy, and we all mindlessly follow this path. That's where all of my credit card debt came from. I was doing it because that's what everybody else does and that's what we're all supposed to do. It was this incredible feeling of freedom and of empowerment when I realized that that was all kind of a big lie, basically, and it wasn't making me happy. I have lots of friends who have big houses and big cars, and they still have a lot of turmoil going on in their lives. That's not the way to find happiness. It's true what you said: we're constantly being bombarded that we need to have more stuff and buy more stuff, and at some point, you have to revolt against that.

Chris: *Yeah, a big part of that messaging always revolves around this notion of happiness, when in reality we're just marketed pleasure. Pleasure is temporary. Happiness is a more long-term notion. So what's your strategy that people should get started on?*

Jenn: I feel like there's not a one-size-fits-all strategy for everybody. The first step that I always tell people is to give yourself permission, which sounds kind of silly. People just want to jump over that and be like, "No, I want you to tell me physically, where do I start, what do I do?" But you have to do this other step first where you give yourself permission, because if you just dive into a pile of stuff and you haven't really clarified why you're doing it, then it's not going to go as successfully as if you take the time to really sit down and figure out why you want to

do this in the first place. What is your end goal, what do you want your life to look like? Could it be a tiny house? It could be having less stuff than in your current house. Whatever it is, it's going to be your end goal that's going to really motivate you and drive you to get there. And then you have to give yourself permission to get rid of your excess stuff. Don't listen to those voices in your head that remind you you've been holding onto Grandma's quilt for 25 years because it was Grandma's—you hate it, and you don't use it. You have to give yourself permission to override that voice and be able to get rid of these things. I also always tell people to make sure that they work in small chunks. Don't stand back and look at your entire house and get overwhelmed and be like, "Oh my gosh, I can never do this, I have too much stuff." Just look at one room at a time, or look at one section of a room at a time. That's going to start to give you that adrenaline rush and that momentum. For some reason, I always feel like the closet is a really easy, nonintimidating place to start.

FINAL THOUGHTS

We all have different motivations and different priorities. Whatever your interest in tiny living is, or in anything else for that matter, pursue it. I wound up building tiny houses quite by accident. But there were so many places along the way where I could have veered off the path, where I could have procrastinated my way into not taking action. Sometimes we let uncertainty and fear get the better of us. In the words of Steve Perry, "Don't stop believing." Keep moving forward, even if you are only taking small steps. And thanks for reading this book. I hope that it helps you move further. I would love to hear from you, so please follow me and reach out via any of the following platforms:

- *www.tinyindustrial.com*
- *www.facebook.com/tinyindustrial*
- *www.instagram.com/tinyindustrial*
- *www.etsy.com/shop/tinyindustrial*

There is a quote that I put up in my tiny house that still inspires me to keep going: "Someday is not a day of the week." Here's wishing you all the motivation you need, too. And most of all . . . enjoy your tiny house!

Tiny House Checklist

	PLANNING
	I know why I want to go tiny
	I'm convinced that tiny living is right for me
	I'm ready at this point in my life to go tiny
	I can handle downsizing stuff as much as will be required to go tiny
	I know what's important to me
	I have a clear vision of what tiny living looks like for me
	I have a timeline for when and how I want to go tiny
	I know what features my tiny house should have and what it should look like
	I'm willing to make tradeoffs between the things I want and need
	I have a grasp on how much I will be moving the tiny house around
	I know if I want to build on a foundation or on a trailer
	I have made a decision on loft sleeping vs. a ground-level bedroom
	I have a plan for how many windows there should be and where they should go
	I have ideas on how the interior space can be expanded to incorporate the outdoors
	I know how I'm going to finance the house
	I know if I'll be doing the construction or hiring someone else to do it

SELF BUILD

	I have access to the tools I'll need to build the house
	I'm comfortable using the tools that will be needed to do the work
	I have a place where I can build the house
	I feel comfortable installing all the systems that the house will need
	I can find people willing to help me complete the parts of the build I don't feel comfortable doing myself

BUILDER BUILD

	I feel comfortable hiring a contractor to build or help me build the house
	I know the level of involvement that I want to have if someone else is doing the work
	I understand that the house will be more expensive if I hire a builder

LIVING TINY

	I have done my research and know where and how I will live in my house legally
	I have a good grasp of all the nuances of living tiny
	I understand how to move/tow the tiny house safely
	I know that some things will be more difficult in a tiny house
	I see more benefits than downsides to living tiny
	I recognize that I am an individual and have decided that this is what I want to do; I don't care if people think I'm crazy or foolish to pursue this path

Resources

Legalities/Informational

www.americantinyhouseassociation.org
www.unitedtinyhouse.com

Blogs

www.tinyhousetalk.com
www.tinyhouseexpedition.com
www.tinyhousegiantjourney.com
www.tinyhouseblog.com
www.relaxshacks.com
www.livingbiginatinyhouse.com

Plans

www.tinyhousedesign.com
www.tinyhouseplans.com
www.padtinyhouses.com

Inspiration

www.reddit.com/r/TinyHouses
www.tinyhousesandbeyond.com
www.tinyhouselistings.com

Shameless Plugs

www.tinyindustrial.com
www.tinyhouseinthecountry.com

Books

Diedricksen, Derek "Deek." *Humble Homes, Simple Shacks, Cozy Cottages, Ramshackle Retreats, Funky Forts: And Whatever The Heck Else We Could Squeeze In Here.* Lyons Press, 2012.

Diedricksen, Derek "Deek." *Micro Living: 40 Innovative Tiny Houses Equipped for Full-Time Living, in 400 Square Feet or Less.* Storey Publishing, 2018.

Diedricksen, Derek "Deek." *Microshelters: 59 Creative Cabins, Tiny Houses, Tree Houses, and Other Small Structures.* Storey Publishing, 2015.

Shafer, Jay. *Jay Shafer's DIY Book of Backyard Sheds & Tiny Houses: Build Your Own Guest Cottage, Writing Studio, Home Office, Craft Workshop, or Personal Retreat.* Fox Chapel Publishing, 2013.

Shafer, Jay. *The Small House Book.* Tumbleweed Tiny House Company, 2009.

About the Author

Photo credit: Mia Fitzmaurice

Chris Schapdick spent decades in a lucrative career in the online media sector working in various management capacities for publicly traded corporations as well as small startups. But professional success felt hollow and failed to provide him with real purpose and deeper meaning. In 2016, Chris bid his corporate career farewell and decided to strike out on his own, launching several businesses designed around building tiny houses and coaching others to follow their dreams of a simpler lifestyle. Chris now splits his time between northern New Jersey, where he lives with his teenage daughter, and the Catskills region of New York State, where he owns and manages his businesses. It is in the solitude of the forest and rolling hills that he finds the inspiration and creative energy that fuel his designs and projects. Chris is a sought-after blogger and speaker on the tiny house circuit and has garnered awards for his builds, including "Best Tiny House in New Jersey" at the United Tiny House Association's New Jersey Tiny House Festival (2017) as well as in November at the same organization's event in Florida.

Photo Credits

All images by the author unless otherwise noted alongside the photo or below.

Floor plans on pages 45, 120, and 141 illustrated by David Fisk.

The following images are credited to Shutterstock.com and their respective creators: cover top: Ariel Celeste Photography; cover bottom left: ppa; cover bottom middle: Lowphoto; cover bottom right: ppa; back cover: Ariel Celeste Photography; 1: Segen; 2–3: Ariel Celeste Photography; 5: ppa; 26: ppa; 27 top left: Ariel Celeste Photography; 27 middle left: The Adaptive; 27 bottom left: Lowphoto; 27 right: ppa; 28 top: Lemon Tree Images; 33 top: Zack Frank; 37: Wyoming Nomad; 38 bottom: Ariel Celeste Photography; 41: ppa; 43 right: oddech; 54 bottom: Mike Focus; 68 top: stocksolutions; 72 top: Thoranin Nokyoo; 79 top: Andrey Burstein; 81: Syda Productions; 83 bottom: Prarinya; 84: Barry Blackburn; 85: ArtCookStudio; 91: Steven Belanger; 97 bottom left: Caron Watson; 106 top: David Papazian; 107 bottom: Igor Meshkov; 109 middle: Alexandr III; 124 bottom: Nakantee; 126 top right: Lighttraveler; 126 bottom: Suwan Banjongpian; 128: Victoria 1; 133 bottom: MoreGallery; 137 bottom: Kucher Serhii; 145 right: Vorm in Beeld; 146: Julie Flavin Photography; 147 top: Artazum; 147 bottom: mubus7; 148 bottom: Lowphoto; 150: Helen Filatova; 162: Ariel Celeste Photography; 170: Ariel Celeste Photography; 174–175: Ariel Celeste Photography

Index

A

adjusting to tiny living, 42
    age of homeowners, appeal of
      tiny houses and, 36–37
air compressor, 88
air conditioning. *See* heating and
    air conditioning
air inlet, passive, 105, 128
American Tiny House Association,
    145
appeal of tiny houses, 36–38
appropriateness of tiny living, self-
    assessment, 39–41
assessing fit for you, 39–41
axles, 59

B

barbecue, backyard, 136
bathroom
    showers and, 72, 103
    venting, 127–28
    water for. *See* plumbing
batteries, deep-cycle, 97–98
batteries, solar, 96–97
Baxter, Jenn, interview with, 168–69
benefits of tiny houses, 36–38
Bennett, Andrew, interview with,
    117–18
bilge fan, 128
black water, 102
box truck tiny house, 44–47
brakes and brake controller, 60, 139
building tiny home. *See also*
    construction; design; electrical;
    finishing touches; insulation;
    plumbing; roofs/roofing
      checklist, 172–73
      deciding who will build, 53
      other resources for, 81, 149–50,
        170, 174
      timeline for home and, 53
Burns, Andrea, interview with,
    73–75

C

campgrounds, parking in, 148
carbon footprint, lowering, 38, 56

checklist, 172–73
children, lofts and, 64
church, tiny, 161
circular saw, 86
classification of tiny houses, 146
clothing, downsizing and, 43
clutter
    American consumerism,
      minimalism and, 42
    avoiding/eliminating, 43, 112
    clothing, 43
    downsizing and, 43
    kitchen, 43
    living without, 43
    media items, 43
colors, tips on, 113
composting toilets, 132–33
connection to the world, your
    desires, 51
construction. *See also* electrical;
    finishing touches; insulation;
    plumbing; roofs/roofing
      about: overview of, 76–77
      doing it yourself (or not), 80–82
      flooring, 91
      framing (exterior), 91–93
      framing (interior), 94
      hiring contractor for, 83–84
      involvement in process, 91
      knowledge and ability to
        perform, 81, 82
      order of, 91–94
      partial sourcing, 82
      research, planning, and
        exposure before, 162–66
      resources, 81
      sheathing, 93
      tiny house manufacturers for, 84
      tools for, 85–90. *See also*
        specific tools
contractor, hiring, 83–84
Craft & Sprout, 51, 52
crafting space, tiny, 159

D

design, 48–72. *See also* finishing

touches; lofts; windows
    about: overview of, 48–49
    checklist to help with, 172–73
    connectedness to the world
      and, 51
    floor plan examples. *See* floor
      plans
    keeping open mind, 53
    mobility and, 51–52, 54–55
    next steps after reading this
      book, 162–67
    off-grid living and, 55–56
    outdoor space and, 70–72
    priortizing desires, 50–52
    research, planning, and
      exposure, 162–66
    roughing out plan, 53
    size and, 51–52
    starting with a vision, 50–52
    timeline for home and, 53
    tradeoffs, 51
    trailer foundations. *See* trailers
dimensions of home, 54
door, insulated, 104
dormers, 63
drill with bits/attachments, 86. *See*
    also impact driver
dry flush toilets, 134
Dufault, Jesse, interview with,
    142–43

E

Eaves, Alex, interview with, 44–47
electric brakes, 60, 139
electric water heater, 103
electrical. *See also* energy
    110-volt AC and 12-volt DC,
      95–96, 99
    batteries, 96–98, 99
    cables and amp ratings, 95
    generator, 96
    hydropower, 98–99
    installation overview, 94
    off-grid living, 55–56
    off-grid power options, 96–99
    on-grid system, 95–96

panel for controlling, 95–96
propane supplementation, 99
sequence in construction, 91, 94
solar power (batteries, cells,
    controllers, panels), 96–98, 99
wind power, 98
electronics, media clutter, 43
emergency exits, 64, 69
energy. *See also* electrical
    cost savings, 56
    independence, 56
    off-grid living, 55–56
    passive house and, 56
entertaining, 135–36
Eschenheimer, Nora, interview
    with, 142–43
event space, tiny, 161
exercise space, tiny, 159
exposure, planning, and research,
    162–66. *See also* checklist; design
exterior of house, entertaining and,
    135–36. *See also* outdoor space

F

fans and venting, 127–28
faux finishes, 111
finances
    about: overview of, 78
    assessing yours, 40
    benefits of tiny houses, 38
    friends, family and, 79
    off-grid living savings, 56
    paying for home, 78–79
    saving up money, 79
    sources of money (loans/
      financing, etc.), 78–79
    spreading costs by building in
      stages, 79
finishing touches, 111–16
    adhering to theme, 112
    avoiding clutter, 112
    color considerations, 113
    faux finishes, 111
    lighting, 114
    proportions, 116
    quality materials for, 113

Notes

Notes

Additional Praise

"Chris offers practical encouragement for the potential tiny house builder and dweller by breaking down all the essential considerations into small, manageable steps."

—Alexis Stephens, co-founder of *Tiny House Expedition* and creator of the *Living Tiny Legally* docu-series

"There are a lot of talented tiny house builders out there, but there are very few with the talent, quality, and diversity of designs that equal the creations of Chris Schapdick."

—John Kernohan, Chairman & Founder, United Tiny House Association

"Chris writes with warmth and humor to help you clarify your tiny house dream and get you going on designing the home and life that is right for you."

—Carmen Shenk, author of *Kitchen Simplicity* and *TinyHouseFoodie.com*

"If this book had been around in 2009, I would have done things differently on my own build and probably had a lot less trial and error with a whole lot more cool factor!"

—Andrew M. Odom, founder of Tiny r(E)volution